FIRST CONTACT

Scientific Breakthroughs in the Hunt *for* Life Beyond Earth

MARC KAUFMAN

Simon & Schuster
New York London Toronto Sydney

Simon & Schuster
1230 Avenue of the Americas
New York, NY 10020

First Simon & Schuster hardcover edition April 2011

Designed by Nancy Singer

Manufactured in the United States of America

10 9 8 7 6 5 4 3 2 1

Library of Congress Cataloging-in-Publication Data

Kaufman, Marc.
First contact : scientific breakthroughs in the hunt for life beyond Earth / Marc Kaufman.—1st Simon & Schuster hardcover ed.
p. cm.
1. Life on other planets. 2. Habitable planets. I. Title.
QB54.K38 2011
576.8'39—dc22 2010044630
ISBN 978-1-4391-0900-7
ISBN 978-1-4391-3030-8 (ebook)

To Lynn Litterine, the brightest star in this man's galaxy.

CONTENTS

FIRST CONTACT

1 THE BIGGEST DISCOVERY OF THEM ALL

If it's just us in this universe, what a terrible waste of space.

But it's not. Before the end of this century, and perhaps much sooner than that, scientists will determine that life exists elsewhere in the universe. This book is about how they're going to get there. And when they do, that discovery will rival the immensity of those that launched our previous scientific revolutions and, in the process, defined our humanity. Copernicus and Galileo told us we were not, after all, at the center of the universe, and their ideas fathered a scientific astronomy that, four hundred years later, is allowing us to be a space-faring planet. Charles Darwin gave us our evolutionary roots, which, a century later, propelled Louis and Mary Leakey on a thirty-year search culminating in the recovery of fossil hominid remains almost two million years old in Tanzania's Olduvai Gorge—proof that humankind began in Africa. So here we are now, the descendants of the skilled toolmakers and explorers who left the continent some sixty to seventy thousand years ago. We've populated the globe and sent astronauts to the moon. Next up: Life beyond Earth.

For thousands of years, humans have wondered about who and what might be living beyond the confines of our planet: gods, beneficent or angry, a heaven full of sinners long forgiven, creatures as large and strange as our imagination. Scientists now are on the cusp of bringing those musings back to Earth and recasting our humanity yet again. "Astrobiology" is the name of their young but fast-growing field, which immodestly seeks to identify life throughout the universe, partly by determining how it began on our planet. The men and women of astrobiology—an iconoclastic lot, quite unlike the caricatures of geeks in white lab coats or UFO-crazed conspiracy theorists—are driven by a confidence that extraterrestrial creatures are there to be found, if only we learn how to find them. Most hold the conviction that if a form of independently evolved life, even the tiniest microbe, is detected below the surface of Mars or Europa, or other moons of Jupiter or Saturn, then the odds that life does exist elsewhere in our galaxy and potentially in billions of others shoot up dramatically. A solar system that produces one genesis—ours—might be an anomaly. A single solar system that produces two or more geneses tells us that life can begin and evolve whenever and wherever conditions allow, and that extraterrestrial life may well be an intergalactic commonplace.

With goals so enormous and compelling, astrobiology has brought forth a new generation of outside-the-box researchers, field scientists, adventurers, and thinkers—part Carl Sagan, part Indiana Jones, part Watson and Crick, part *CSI: Mars.* They are men and women who drop deep below the surface of the Earth or tunnel into Antarctic glaciers in search of life in the most extreme places, who probe volcanoes for clues into how Earthly life began, who propel life-detecting robots and ultimately themselves into space. They come up with ever more ingenious methods for detecting planets that circle distant suns; they scour our planet for Mars- or Europa-like habitats they can minutely study for the life-supporting conditions they might encounter when our spaceships arrive there. They probe the cosmos as far as 13 billion light-years away for signs of the earliest stirrings of the order and chemistry that created life on Earth. Some are even working to define and understand "life" by creating it in the lab. They've harnessed that childhood excitement so many of us felt when, on hot, hard-to-sleep summer nights, we tried to imagine what it would be like to visit Mars (very dry), or travel to the end of the universe (very confusing), be around when life first began (very lonely), or come across extraterrestrial life (very exciting). The world has changed enough that today, a large and growing number of scientists are earning their livelihoods turning their imaginings into hypotheses and putting them to tests inconceivable even a decade ago.

Why now? Why does the promise of cracking the extraterrestrial barrier seem close enough that so many prominent scientists from NASA to the Massachusetts Institute of Technology, from the Carnegie Institution of Washington to Princeton and Cambridge universities, have decided to ignore the giggle factor associated with UFOs and *ET* and join the quest? The answer, put broadly, is that the field is getting results.

In the past ten years we have found that hundreds of planets orbit distant suns not too different from our own and can reasonably infer that billions more exist. Many are bound to be rocky planets in eminently habitable zones the right distance from stable suns to give life a chance. More than five hundred of these exoplanets ("exo" because they orbit suns other than our own) have already been identified and even more new ones are being discovered every week. In the past two decades, we have also explored a vast world of microbial "extremophiles" that live in Earthly environments once assumed to be incapable of supporting life—findings that make it easier to hypothesize that life survives in "uninhabitable" conditions on other planets and moons, too.

Extremophile research started with microbes living in hot springs like Yellowstone and near deep underseas "black smoker" thermal vents that are even hotter. Each year scientists reach further and almost always get results—finding life miles underground, encased in ice, or bathed in acid. A very different group of researchers is also getting closer to synthesizing something akin to life in the lab, research that sets the stage for an understanding of how life might have started on Earth and elsewhere. One of those labs will soon have produced self-replicating genetic material out of nonliving component parts—in other words, created something very life-like from synthesized genetic material. And planetary scientists are finding ever more reason to conclude that Mars in particular—written off as lifeless thirty years ago, after NASA's Viking missions—has, or had in the past, most everything necessary to support life: liquid water, carbon compounds, nutrients, and a minimally protective atmosphere. In 2009, the life-on-Mars theory got a major boost with the confirmed discovery of methane gas in its atmosphere. On Earth, 90 percent of methane is produced through biology.

The field of astrobiology in its modern form came into existence in the late 1990s, following an announcement by NASA that its researchers had found likely signatures of life in an ancient Martian meteorite that landed in Antarctica. The proof supporting that conclusion was contested by many scientists, but the study of meteorites from Mars and elsewhere has blossomed anyway. Since then, increasingly sophisticated instruments have allowed researchers to tease more widely accepted secrets from the rocks. All these very concrete discoveries—and the fact that interstellar space is full of potentially life-supporting, carbon-based compounds that constantly rain down on us and on other celestial bodies—have convinced scientists around the world that it's highly unlikely that Earth is the only place in the universe where life arose.

This is heady stuff, and it has scientists moving in all directions. If primitive life forms can exist miles below the Earth's surface in the mines of South Africa without contact with the sun or its products, why couldn't the same be true on Mars, or on the moons of Jupiter and Saturn, or on the untold number of rocky planets we now know exist across the universe? The same logic applies to extremophiles in the abyss of the deep sea, in glaciers tens of thousands of years old, in Spain's acidic Rio Tinto or California's alkaline and arsenic-laced Mono Lake. Extremophiles even survive in the upper atmosphere of our planet.

Perhaps you're thinking that the discovery of bacteria deep underground does not seem all that earth-shattering. Perhaps you're thinking, too, that even if Mars or a moon in our solar system turns out to have comparable subterranean life, that would prove little except that primitive life forms come into being and survive in all kinds of places. Astrobiologists see things very differently. As they are quick to point out, life has existed on Earth for at least 3.8 billion years, and for more than 3 billion of those years single-celled bacteria and related microbes were the only living things around. In other words, butterflies, tree sloths, saber-toothed tigers, and humans all evolved from single-celled organisms too small to see without a microscope. Astrobiologists today have a deep respect for the significance of bacteria and other single-celled creatures—and their ability to evolve into intelligent life.

Searching for and understanding extremophiles is almost universally embraced by the scientific community as essential and revelatory science now, but as late as the mid-1990s it was seen as quixotic and something of a career ender. Tullis Onstott is the man who changed the field by descending into deep gold mines in South Africa and coming back with remains of bacteria that have lived down there in their own peculiar worlds for millions of years. Onstott, a geobiologist at Princeton University, initially couldn't get funding for his research, and his first expeditions were paid for out of his own pocket.

Stories like his are common in astrobiology. The early extrasolar planet hunters were told in the 1980s and early 1990s that they were wasting their time, that there was no way to detect their quarry through the blinding glare of parent stars hundreds or thousands of light-years away. So they developed other techniques based on measuring the slightest movements of those suns, minuscule course corrections caused by the gravity of the orbiting exoplanets. Now, through those methods, planets beyond our solar system are found regularly and the expectation is that billions more remain to be mapped. What we know of them remains limited to their orbits, their mass, and a little about their component parts. The new challenge is to characterize them much better, especially the smaller Earthlike planets expected to be discovered in the years ahead. But these planets are minute at such great distances and are blotted out by the intense light from their parent stars.

So how can astronomers compensate? One proposal, years in the making, involves sending into deep space a football-field-sized sunshade, that would then work in tandem with an orbiting telescope 35,000 to 50,000 miles away to create an "occulter." The flower-petal-shaped screen would block light from the star and thereby allow the telescope to see and study orbiting planets, their atmospheres, and any signatures of possible life. The long process of transforming an idea like this into a space-faring reality got a big boost in 2010 when a panel of the National Academy of Sciences gave its highest priority to exploring exoplanets and their atmospheres in the next decade. An occulter system may ultimately not be the technology selected for the job, but it is a serious contender.

The sunshade might sound like a far-fetched project to pursue, but many of the most successful results in astrobiology began as pursuits that sounded impractical or extreme. Take, for instance, the work of Sara Seager, the astrophysicist who first opened my eyes to the breakthroughs and great promise of astrobiology. Her mind easily visits places where few of us can follow. Raised in Toronto, she puzzled her father with an early interest in outer space, which later turned into degrees in math, physics, and astronomy. Her pioneering work on the atmospheres of exoplanets is what persuaded MIT to offer her tenure and an endowed chair in planetary science. She was thirty-four at the time. She is a theorist rather than a hands-on planet hunter: her scientific specialty is to predict and refine ways to identify the elements and compounds in the atmospheres of extrasolar planets, work that she began before the first extrasolar planet was detected. She was told when she started that her ideas were theoretically interesting but couldn't be tested in her lifetime. But little more than a decade later, we know that the gas methane exists on a giant planet orbiting a star 63 light-years away, that sodium exists on a planet orbiting a sunlike star 150 light-years away, and that evidence of both oxygen and carbon has been detected enveloping another planet in that solar system. Now Seager is convinced that extraterrestrial life will be detected within her lifetime, and she wants to be part of that triumph.

Given the size of the challenges taken on by astrobiology, however, the big break won't come from the inspired minds of one or two great thinkers, as they did with Galileo, Copernicus, Newton, Darwin, Einstein, and Watson and Crick. This search is more of a broad-based, inexorable, but oddly unheralded Apollo program, an undertaking that requires thousands of researchers with very different backgrounds, technical skills, and obsessions. The enterprise is playing out in plain view, yet is so big it is almost invisible.

Some astrobiologists (and astrobiology fans) no doubt dream of a "Eureka!" moment when life is discovered beyond Earth or synthesized here—an equivalent to Neil Armstrong's giant step on the moon, or the unraveling of the structure of DNA. Someday that may come, but science generally works incrementally, and takes much smaller bites. Even the biggest, hottest research questions in astrobiology involve work akin to crime-scene forensics, often drawing on small left-behind clues to help put together pieces of the larger puzzle. These are the kinds of questions absorbing, inspiring, and at times dividing the inherently fractious tribe that constitutes the field of astrobiology.

That tribe is fractious because it's attempting to answer a set of unavoidable and obnoxious questions—obnoxious because they appear so simple, yet actually are so complex: What, when all is said and done, is life? Could we encounter it elsewhere and simply pass it by? Are we blinded to extraterrestrial life by our Earth-based assumptions of what life must be? We have a substance on Earth—a blackish rock coating called desert varnish found in many arid places and often used as a background for Native American petroglyphs. Experts in the field going back to Charles Darwin have studied it, and they still sharply disagree about where it comes from: whether it is a product of microbial biology or of geology and chemistry. Getting a better sense of what is living and what is not on Earth seems pretty essential to the quest for life beyond Earth, and so these borderland cases attract lots of attention. Desert varnish is especially intriguing because something that looks similar to it has been seen during several Mars missions, or so some scientists contend.

Nobody knows how or why, but virtually all the amino acids—molecules that make up essential building blocks of proteins (and therefore of life as we know it)—share a necessary quality that is otherwise seldom seen on Earth: Their molecules are all organized in a formation that scientists call "left-handed," enabling them to interact with uniformly "right-handed" sugars. Because virtually nothing else on Earth is structured like this—all "left-handed" or all "right-handed"—some scientists suspect the initial overabundance of left-handed amino acids arrived here by way of meteorites or comet dust. Evidence from one large and quickly recovered meteorite that fell in Australia in 1969 lends some support to that conclusion, with potentially major implications about how life began and evolved here, and the possible makings of life elsewhere.

Did life on Earth start in scalding, sulfurous hydrothermal vents on the deep ocean floor, or perhaps at the less intense side vents that tend to spring up in the same regions? Did it start in subterranean rock fractures where it could be protected from the heavy meteor bombardment of early Earth? Did it begin around the plumes of erupting volcanoes, where intense lightning activity (the kind of energy needed to start the chemistry needed to support life) is now known to be common? Or did it begin in the "little warm ponds" put forward by Charles Darwin, or perhaps via those meteorites? All of these possibilities have their advocates. If scientists can get a clear sense of how nonliving chemicals were transformed into the self-replicating, energy-consuming, evolving entities that ultimately produced us here, they'll have the beginning of a road map for what might be happening out there.

While this search is under way, another hunt for signs of extraterrestrial life—a broad range of potential "biosignatures" ranging from the presence of liquid water to the organic molecules associated with life on Earth—is also moving quickly ahead. For instance, the methane gas recently detected in the atmosphere of Mars is released in plumes at specific sites and at predictable times, suggesting previously undetected, even unimagined Martian geology and biology. On Earth, about 90 percent of methane is a by-product of biological processes released by living, or once-living, things ranging from bacteria to rotting trees to flatulent cows. That isn't necessarily the case on Mars, but it certainly is a real possibility.

Scientists are also probing whether Mars was more hospitable to life at its inception than was Earth, which after all did take quite a hit when a Mars-sized body crashed into it and ejected the material that most planetary scientists believe became the moon. And if life did start on Mars, could it have traveled via ejected rock-turned-asteroid to Earth? Bacteria in Antarctica and other glaciers frozen for hundreds of thousands of years come back to discernible life when brought to higher temperatures, and researchers contend they could last in a suspended state (or maybe even carrying on life functions) for millions of years more. Other microbes have shown a previously unimaginable ability to withstand the cosmic radiation of space. Put all this together and the unavoidable question becomes whether, at bottom, we're all Martians—quite literally descendants of life from Mars. If methane can ultimately be traced to a biological source on Mars, astrobiology will enter an entirely new phase and the quest to find extraterrestrial life will become something more like a race.

Have we actually already found extraterrestrial life on previous Mars missions and in meteorites found on Earth? This is one of the most contentious issues in astrobiology—and in science as a whole—and many highly qualified scientists on opposing sides of the issue are 100 percent convinced they're right. Feelings are especially high because as astrobiology's patron saint, Carl Sagan, once said, "Extraordinary claims require extraordinary proof." But extraordinary proof is very hard to come by, and tantalizing findings are hard to keep under wraps. The result has been a number of long-running scientific grudge matches—intellectual blood sport at the highest of levels, with seemingly many rounds to go. Interdisciplinary cooperation is the mantra of astrobiology but it has yet to repeal the laws of human nature.

The most significant dispute is no doubt over the contested discovery that gave birth to the new era of astrobiology: the 1995 announcement that NASA scientists had discovered a meteorite from Mars that contained numerous features consistent with extraterrestrial life. Critics quickly tore into the report and left it seriously wounded. But the authors have continued their work and say they are more convinced than ever that many Martian meteorites show signs of long-ago extraterrestrial life.

Just as those immersed in astrobiology now theorize that extraterrestrial life does—perhaps even must—exist, astronomers long theorized that planets circled stars in other solar systems. It wasn't until the mid-1990s, however, that the first definitive detections were made. Now, more than five hundred exoplanets have been identified, seven hundred more are awaiting confirmation, and billions more are believed to exist throughout the universe. As much as any other discoveries, the peek into the world of exoplanets has supercharged astrobiology and encouraged scientists to substantially increase their bets on the existence of extraterrestrial life. But the discoveries have come with big surprises. Most of the extrasolar planets found so far are large gas giants like Jupiter, orbiting close to their suns with smaller but also giant planets farther out—a kind of solar system that virtually nobody predicted. That so many of the planets discovered are in this category is, to a substantial extent, a function of how astronomers are looking for them—bigger and closer to the central star is what we have the technology to detect. But the notion that any Jupiter-sized planets would be orbiting their suns in four or five days was, until recently, unthinkable. Equally unexpected was the discovery that many solar systems consist of planets that travel in wildly eccentric orbits, not the circular or near-circular ones we're accustomed to. The fact that solar systems come in such peculiar arrangements has both promising implications for astrobiology—with solar systems so varied, the probability is that some others are "just right"—and some negative because planets in those wildly eccentric orbits would probably make their solar systems unstable and uninhabitable.

So the big question for planet hunters is no longer simply how to find planets, but rather how to find more of the smaller, rocky, Earth-sized planets the right distance from their suns to be potentially habitable, and to find solar systems structured in ways that could allow these cousins of the Earth to become nurseries for life. NASA's Kepler spacecraft was launched in 2009 to make a broad search for Earth-sized planets, and it's expected to begin delivering substantial results in 2011. But much of the serious planet hunting is being done using Earth-based telescopes, and the ingenuity of the scientists operating them is the stuff of legend. Anyone betting against them finding habitable planets and solar systems has not been following their fevered discoveries.

Astrobiologists are constantly searching for habitats on Earth that can be studied as near cousins to environments that might be found on other planets or moons—the parched Atacama Desert in Chile, the hydrothermal vents of Yellowstone Park and the ocean floor, the dry valleys and deep glaciers of Antarctica. One of the more compelling sites is Lake Bonney in Antarctica, which has a deep covering of ice over liquid water known to support microbial slimes and life. Jupiter's moon Europa also has a thick layer of ice over what is now believed to be a vast ocean of liquid and perhaps life-supporting water, and NASA and the European Space Agency have proposed it as a major "flagship" mission for the 2020 time period. NASA believes Lake Bonney can serve as a useful analogue to Europa for research purposes, and so it is testing sophisticated submarine vehicles there—autonomous robots whose offspring may well find themselves someday on that moon's icy surface.

But the study of habitats has a more cosmic meaning, too. Solar systems are now described as having (or not having) "habitable zones"—regions where rocky planets with atmospheres could exist, and where the sun heats the planet to the right temperature for liquid water. Since we now know that the complex carbon-based organic materials that are the building blocks of life on Earth can be found throughout the universe—that they fall on exoplanets just as they fall on Earth—it seems quite unlikely that life wouldn't start and evolve quickly on an otherwise habitable planet. This, of course, is based on the presumed dynamics of early Earth, our one and only example of a life-supporting planet. Earth is known to have formed about 4.5 billion years ago and to have undergone hundreds of millions of years of meteorite bombardment and generally hellish conditions. Yet early forms of life have been traced as far back as 3.8 billion years on Earth, suggesting that life arose not too long, in geological terms, after conditions became favorable.

In trying to define what makes life possible, astrobiologists are forced to confront another question: Is life inevitable, or the result of a series of accidents? Did the universe have to be finely tuned to make it possible? This is an unavoidable question because the slightest change in many of the basic physical and cosmological laws of the universe would make it an entirely inhospitable place. A minute increase in the extreme weakness of gravity, for instance, would make stars like our sun burn out in 10,000 years instead of 10 billion. If the neutrons found in every atom were not .01 percent heavier than protons found in every atom, then the universe would allow for no chemical reactions because all atoms would be stable and unchanging. Is this kind of "fine-tuning" a coincidence of almost unimaginable proportions? Does it mean the universe itself is the product of a sort of Darwinian evolution? Does it mean there are many, perhaps an infinite number, of other universes that are not organized in a way that can support life, leaving us by definition in the one that can? Or, leaving the realm of science for a moment, is this "fine-tuning" a cosmic reality that supports the argument for a Creator?

The broad-based effort to answer these and many other questions is remarkable because it has finally made it legitimate for white-coated scientists (actually, mostly the blue-jeaned kind) to spend their careers studying the possibilities, locations, and signatures of alien life. Astrobiology projects now attract more grant proposals from members of the National Academy of Sciences (who are invited to join because of their accomplishments and prominence) than any other subject at NASA. In astrobiology today there's no talk of UFOs, no wormholes or time travel, no giant "gas-bag" creatures floating through the upper reaches of Jupiter (as imagined by Sagan himself). Rather, it's about hard-core science that, until recent years, was technically impossible or simply unimagined, and it stretches from the bottom of deep Earthly mines to the farthest reaches of the universe with its 100,000,000,000,000,000,000,000 (or more) stars, and their unfathomably large number and variety of planets and moons. Even the search for extraterrestrial intelligence, or SETI, has become much more scientifically sophisticated—enough so that NASA and the National Science Foundation have reopened their grant competition to SETI projects, and Microsoft cofounder Paul Allen donated $25 million to begin construction of an array of 350 radio telescopes in northern California designed in part to pick up transmissions from distant civilizations.

NASA has embraced this search, but quietly. Unlike the Apollo missions to the moon, construction of the international space station, or the George W. Bush administration's proposals to settle astronauts on the moon and send them to Mars, no big announcement was ever made about a new NASA push to find extraterrestrial life—and that's probably politically astute. Imagine the chuckling and high dudgeon in Congress had it received an expensive and dicey proposal to find ET. A NASA vision statement released in 2002 made this emphasis on astrobiology explicit, declaring the agency's goals thus: "To improve life here; To extend life to there; To find life beyond." By 2006, all reference to finding "life beyond" had been removed, but the goal had already been hardwired into the actual workings of the agency.

The NASA astrobiology program was formally initiated late in the Clinton years with a modest budget and a small bureaucracy of its own. The agency's Astrobiology Institute gives out modest but still very competitive grants totaling about $50 million each year. But that's only the most obvious effort. NASA and European Space Agency missions are regularly designed with extraterrestrial life in mind. The most eagerly anticipated include the Mars Science Laboratory (designed to scour Mars for signs of the chemical building blocks that make life possible), two joint NASA-ESA missions to Mars (inspired and configured, to a significant extent, by the discovery of methane on the planet), and an increasingly possible NASA-ESA mission to Europa. Then there's the biggest prize on the horizon—a mission to Mars to gather up rocks and soil and bring them back to Earth for the kind of exhaustive analysis scientists have dreamed of for decades.

As science has found Earth to be a mere speck in the universe, the notion of our human specialness has diminished—perhaps one reason why the centuries-old debate about the existence of extraterrestrial life has at times been so raw. When the discovery of extraterrestrial life comes, the process begun by Copernicus and Galileo in the sixteenth century of pushing the Earth away from the assumed center of the universe will have come full circle. But there is also the strong possibility that astrobiology will introduce people to a transformed understanding of the cosmos and our place in it. That's what Steven J. Dick thinks. He's a trained astrophysicist who served for many years at the U.S. Naval Observatory and later as NASA's chief historian, and is the author of numerous books about the history of thinking about extraterrestrial life. "With due respect for present religious traditions whose history stretches back nearly four millennia," he suggests, "the natural God of cosmic evolution and the biological universe, *not* the supernatural God of the ancient Near East, may be the God of the next millennium. . . . As we learn more about our place in the universe, and as we physically move away from our home planet, our cosmic consciousness will only increase." Because what, in the end, is nature?

If life is found on Mars or Europa, then isn't that nature as well? And if carbon-based organic material fills significant portions of space, and is found in meteorites broken free from planets, comets, and asteroids, then isn't that nature, too? Feeling at home in nature suddenly has a very different, much bigger meaning. That, really, is what astrobiology wants us to understand: that the universe we're privileged to inhabit is more complex, more fertile, and more mysteriously grand—yet also more knowable—than we could possibly imagine.

2 REALLY EXTREME LIFE

Science moves ahead on hunches. Tullis Onstott, the Princeton University geobiologist, first descended into a South African gold mine on a hunch in 1996, using six thousand dollars of his own money and carrying, instead of the usual pickaxes and dynamite, a small hammer, a chisel, some vials for collecting water, and some sterilized bags for collecting rocks. Over the next decade, he and his fellow mine divers found microbes that broke nearly every rule of life. Up until then, it was taken as scientific fact that to survive, a creature needs an energy source and an environment that isn't extremely hot or cold, isn't overly acidic, alkaline, or salty, isn't suffused with radiation or under great pressure. Creatures also need to reproduce or split with some regularity. On his first trip into the mines Onstott found microbes living as far down as two miles that lost out on virtually all of these counts. His prized discovery, made a few years later, was of a bacteria nourished by food—molecules, actually—freed up using energy released by the radioactive decay of surrounding rocks. The microbe also needs some few minerals to survive and some water, which is hidden away from view until miners open up tunnels and bore holes, tapping into underground lakes, streams, and even tiny fissures within the rocks. Not only do these microbes live and move around miles below the surface, but they seem to split—that is, reproduce—as seldom as once a century.

A reading of the genome of Onstott's star bacteria, as well as analysis of the "age" of the water that is often its home, says that the microbe has not seen the light of day, or interacted with anything produced from sunlight, for 3–40 million years. But it has DNA, reproduces, and is clearly alive. The researchers who sequenced the genome found that it had highly unusual abilities to directly take in needed carbon and nitrogen from nonliving sources—very useful abilities given the absence of carbon-based life in its isolated and unrelentingly harsh environment. It even had genes for a tail of sorts, a whiplike growth that would allow it to swim to hidden sources of nourishment. The bug, Onstott concluded, is widespread in a 130-mile-long subterranean region of the gold belt of South Africa. To honor the creature and the world to which it long ago traveled and made its home, the team sought a name in line with the achievement—first of the bug's existence, and then their discovery of it. They found it in the secret Latin inscription that Professor Von Hardwigg, hero of the Jules Verne classic *A Journey to the Center of the Earth,* comes across at the beginning of the book. The parchment directs him to a volcano in Iceland and tells him: *Descende, audax viator . . . et terrestre centrum attinges* (Descend, bold traveler . . . and you will attain the center of the Earth). And so the world was introduced to *Desulforudis audaxviator,* extremophile par excellence.

South Africa is today the center of his research not because similar microbial life doesn't exist far below New York or London or Tokyo, but simply because this is where the deepest mines have been dug. Onstott had first explored the deep underground for microbes as part of a Department of Energy drilling program in Savannah, Georgia, and later at a Texaco well site in western Virginia. Frustrated by his limited results and fears of contamination in his samples, he cast around for alternatives and landed on South Africa's gold, platinum, and diamond mines—with shafts descending two miles and more. But mine owners were reluctant to let strangers into their domains—so much potentially to lose, so little to gain. It took Onstott and others two years of negotiating to get into the mines and later achieve their breakthroughs. Today, he and Esta van Heerden, the head of the Extreme Biochemistry research group at the University of the Free State in Bloemfontein, have won the confidence of the men who run many of the gold, platinum, and diamond mines of the Witwatersrand Basin, the most productive in the world. When a potentially interesting section of mine is opened, or is going to be shut in forever, the mine operators now call van Heerden to give a heads-up.

Maybe it was to redeem the dark history of those mines—flashpoints during the apartheid era and still controversial because of the pay and inevitably harsh work conditions—that operators took a chance and allowed the scientists in. Or maybe mine officials saw value in having their mines become known for something other than producing shiny metals and wealth at a sometimes high environmental and human cost. In any case, their cooperation has been a godsend to astrobiology and has led Onstott and others to conclude that *D. audaxviator* and untold trillions of other underground microbes also live miles below your shopping center, your bedroom, your favorite national park. Or miles below the surface of Mars, for that matter. Eons ago, our most similar planetary neighbor was far more hospitable to life than Earth, which had endured the collision with a smaller planet that produced the moon. But Mars somehow lost its magnetic field, its atmosphere, and ultimately its ability to hold liquid water on its surface or to protect against solar radiation and deadly ultraviolet light. Mars scientists have long speculated that primitive organisms met the new challenges by descending below the surface and adapting through a desperate evolution. Now living proof exists of a potentially parallel scenario on Earth.

I wanted to see this proof for myself, or at least see its subterranean home. That's how I found myself suiting up one January day in 2009 for a descent into a mine owned by Northam Platinum, a sprawling operation in the northern bush of South Africa, just beyond the aptly named Crocodile River. The daily routine on the surface was ordered, polite, and matter-of-fact. Geologists in their white overalls inspected new equipment; managers made sure the shifts were coming and going as planned. Even the miners, lined up in long rows waiting for the trip down to their work, were smiling and chatting; and some were swaying to the bouncy music coming from the loudspeakers. The grass was clipped, the grounds were clean, the five-thousand-worker plant was humming.

Our group of scientists was outfitted in overalls highlighted with fluorescent striping, heavy rubber boots and goggles, hard hats capped with a miner's light, and a mandatory safety kit strapped to our belts, including a breathing device that can filter out carbon monoxide. The chatter ended as we were ushered into the manager's cage; the miners piled into another. The doors slammed shut and both cages picked up speed, plunging down to Level 7—a thirty-miles-per-hour express ride into the crust of the Earth, accompanied by the sound of falling rocks hitting our carrier as we sped by. We jolted to a stop, an attendant pried the door half open, and we stumbled out into a high-ceilinged, man-made chamber that housed a rail yard for miniature ore trains. Accompanied by an array of mine officials, we passed through this small island of light and set off for the outer reaches of Level 7. We were 1.1 miles below ground and surrounded everywhere by dark, gloomy rock.

The scientists, a South African, a Belgian, and a Spaniard all affiliated with the Extreme Biochemistry group at the University of the Free State in Bloemfontein, some three hundred miles away, have descended many times into the deep underworld in search of extreme forms of life. It remains a daunting—and risky—venture, but the search for extremophiles is sending researchers to scores of equally harsh environments around the world. The results have led to a revolution in thinking about the tenacity and adaptability of life, which has been found to thrive in ice a mile deep, in hot springs, in highly acidic rivers, in and around the scalding thermal vents at the bottom of the ocean. The word *extremophile,* after all, means literally a lover of extreme places. For them, it's our world that's toxic.

At Northam Platinum, each turn on the journey to the Level 7 outskirts led to a smaller, darker, hotter tunnel. Like the strands of a spiderweb, the tunnels radiate out from the center in a mazelike order, logical yet quickly incomprehensible to the unguided. The dry path gave way to sloppy mud from the water seeping through the fissures of the rocks, boot-grabbing stuff that hid the old railway ties and pipes waiting to trip us up. A massive ventilation system worked to keep the heat down, pumping air and sometimes water through wide, striated tubing made of dun-colored fabric. Fastened to the sides of the winding tunnels, they gave new visual meaning to the phrase "bowels of the Earth." But that monumental ventilation effort provided limited relief: Ambient temperatures spike with any significant drop into the Earth, and Northam Platinum has an additional heat load to cope with. The radioactive decay of granite in the Bushveld area has long been especially active, and the result is greater than usual subterranean heat.

Time slipped away in the dark sameness of the march. We passed loudly buzzing ventilation substations, rest areas where sooty and muddied African miners were far less likely to talk or smile than those on ground level, and some tunnel branches sealed up with hazard signs warning of methane gas or rock slides or water. I could tell we were nearing our destination when the temperature spiked. We were in a dark, dead-end area that housed an array of equipment installed to (with only limited success) stop a flow of water into the tunnel. The lights from our miners' hats gave fleeting glimpses of the spectral landscape. All around, the soggy heat had aged the gears, chains, and metal slabs in fast-forward, producing what looked like a very long-ago, sea-bottom shipwreck. Dripping stalactites of calcium carbonate took the place of seaweed; corrosion and rust replaced the barnacles. Water dripped from small fracture holes in the rock, spat out of a corroded valve, and drained into a shin-deep pool. I reached out to touch the spray, and it was bathwater hot.

The most excited man in the darkness was Gaetan Borgonie of the University of Ghent in Belgium, a nematodologist on the verge of a potentially major discovery. He had set out a few years before in search of new varieties of minuscule but ubiquitous and extraordinarily hardy roundworms—which aren't really worms at all but rather are called nematodes.

These creatures, Borgonie was eager to explain to me, are many-celled and have both a nervous system and a digestive tract. They are among the most primitive life forms to have that in-and-out system, a rudimentary nervous system, and the ability to reproduce both sexually and as single-sex hermaphrodites. He thought that finding these complex creatures alive this deep in the netherworld would be the first strong signal that more complex life just might survive alongside single-cell bacteria and other microbes deep below the inhospitable surface of Mars or other celestial bodies—places where the sun also never shines, and where the products of sunshine are similarly completely absent. Nematodes, and even three-foot-long tube worms, were discovered some decades ago at or near the dark bottom of the deep ocean, but the extraterrestrial implications of their discovery were less dramatic because the creatures fed on life produced in collaboration with the sun, nourishment that then made the long drop to the ocean floor.

Borgonie's search had taken him to the tiny Caribbean island of St. Croix, to a sulfur-based ecosystem in Mexico, and finally to the South African–American team that had pioneered deep-mine searches for bacteria. Over the past two years, Borgonie—who as a young man wanted to be an astronaut but now was regularly headed in the opposite direction—had descended into South African mines more than twenty times, and had found (along with his colleagues) an impressive number of nematodes and their eggs in gold, platinum, and diamond mines as far down as 2.2 miles. Most had come out of capped boreholes, but not all: He had also learned on previous samplings that before the minerals harden to produce the long, hard stalactites hanging before us, they sculpt cones that, while still wet, can provide an improbable home to his nematodes and their eggs.

Virtually nobody in the field (including the committee at his university that granted him a sabbatical) believed his deep-mine nematode research would succeed because, on land, the worms are known to live only in the soil and near subsoil. He was eager to test his dissenting views and hopeful that he would confirm that nematodes (like bacteria) can also adapt to life deep underground, and to determine how they got there and whether they've been there long enough to evolve into unique creatures.

"These are good, very good," he said to nobody in particular as he waded back and forth between the stalactites hanging from the spectral ironworks on the walls and the measuring and collecting instruments stored in a beaten-up backpack. He carefully removed several cones and collected the precious drips from others. "This system they've worked out is just beautiful to think about and to see," he said.

It was time to go to our primary destination—another dead-end tunnel a little farther on with a borehole that was running especially hot water. The Northam mine's chief geologist knew about the extremophile hunters from Bloemfontein and called van Heerden to tell her the team might want to come down to take a look before the section was dynamited.

Having left the shipwreck site, I marched behind the others. It was unnerving to be so far underground, so completely surrounded by rock. But the tunnel was straight and solidly dug, the ventilation brought in some fresh air and took out the noxious gases, and the periodic sight and sound of miners down branch tunnels kept things from becoming too otherworldly. I came to a junction, made a sharp left to follow the others, and, with the suddenness of a fast-passing train, was staggered by a blast of heat. I reached for the wall to keep standing. Working in this kind of heat is known to bring on hallucinations, and for the next half hour I periodically did see halos of light far broader than anything coming from my miners' lights. Only later did I learn that both Gaetan and Derek Litthauer, a rugged and experienced professor and mine diver from Bloemfontein, have on occasion been evacuated from an especially hot or airless tunnel and sent back to the surface.

I inched my way to the researchers and mine officials gathered ahead. Already the scientists had attached their equipment to a narrow metal pipe poking out from the rock and were collecting water. It was a borehole, one drilled by miners to see what conditions were like inside the rock. The temperature gauge showed the water was a scalding 150 degrees at the end of the pipe. Mine geologist Werner Lamprecht, clearly proud of the extremity of it all, said the temperature several feet inside the rock face was probably in the range of 170 degrees. Steam danced up from the water pooled on the tunnel floor.

I sat on a discarded board beside the tunnel wall and watched. The tunnel had no insects, no spiders, none of the unexpected movement that comes with creatures. Yet previous expeditions had proven that we were not alone—that even this place somehow supported life in the tiny, watery cracks in the rock face, in the dripping stalactites, and who knows where else. We knew something was there because four years earlier Onstott had already discovered several microbes at or near the bottoms of South African deep gold operations. One, located in 2005 in this same Northam Platinum mine, 1.2 miles down, was a highly unusual "star" bacterium featuring a four-to-nine-point star formation—an adaptation that allows it to capture more of the "food" it needs to survive with increased surface area.

It was hard for the parboiled scientists to stay at the final collection site, but it was also hard to pull away, since they never know when they'll be invited for a return expedition. They had a dozen or more tests to conduct, and a variety of filters to place on the flowing water in the hopes of collecting unusual bacteria, nematodes, or something even more unexpected. But as excitement gave way to exhaustion, they packed up for the hike back to the cages. They knew they had to conserve energy because they would return the next morning to collect their filters with the miners' morning shift at 4 A.M.

There's an inevitable needle-in-a-haystack quality to these subterranean searches—the mines are huge, and the hiding places for microbes and nematodes are small and dispersed—and so Borgonie was not particularly disappointed when initial cultures and examinations in the lab came up empty for nematodes from Northam Platinum. The setting still seemed hospitable for his "worms," especially if his theories were on target about how the nematodes follow their prey (the bacteria) through rock cracks to the deep underground, and then adapt to life there. But nobody strikes gold with every dig. The samplings continued, and the very next week, at the gold mine at Driefontein to the southwest, his fortunes changed dramatically. The catch: four healthy nematodes. In all, he found worms or their eggs in seven of more than twenty deep underground South African mine sites he sampled over six months—a discovery that, if nailed down, would dramatically change our understanding of what can survive in the deep netherworld and how it comes to be down there.

Borgonie left for Belgium soon after to spend some time in his lab in Ghent. He hadn't initially planned to return to South Africa, but the lure of the mines and the creatures they hide quickly pulled him back—especially once he was able to convince Onstott to join him in writing the scientific paper they hoped would introduce the "worms from Hell" to the world. This time Borgonie would place filters on the boreholes and leave them for weeks or months to see what else might be living in the rocks.

Six months later, Borgonie, Onstott, and their colleagues were convinced they had, for the first time, detected the presence of multicelled organisms up to 2.5 miles below the Earth's surface. Several samples had been cultured and had begun to squirm and even reproduce asexually in the lab. The team tested for possible contamination—nematodes brought in on miners' shoes, or deep "old" water that somehow had mixed with water from near the surface—and found that the nematodes coming out of the rock were different from anything found in the tunnels. (They didn't include the nematodes found in the stalactites in the paper because Borgonie concluded their presence raised too many extraneous questions.) What's more, the water used for ventilation and cleaning in the mines contains chemicals that kill bacteria and nematodes—strengthening the case that the creatures Borgonie had found in his filters came out from deep in the rock, and not from a miner's boot. Carbon 14 dating of the "worms" and their remains indicated they had been living in the netherworld at least 4,000 to 12,000 years—less time than the team had expected, but perhaps an insight into what allows the worms to survive and what kills them. The older water—millions of years under the surface—has virtually no available oxygen, while the younger water appears to have just enough to keep the aerobic nematodes alive. The team's excitement was palpable. As Borgonie and Onstott said in a summation of their work, "The presence of nematodes kilometers beneath the surface of the Earth is like finding Moby-Dick in Lake Ontario."

Once the research is published, scientists will pick it apart, assessing whether the data about contamination is convincing and whether nematodes can really live in an environment so hot, so lacking in oxygen, and with so little food. The discovery suggests that nematodes, which are already known to live around deep-ocean hydrothermal vents in temperatures as high as 180 degrees Fahrenheit, probably also live below the ocean floor and should be pursued there. Living below ground and below the ocean floor would give the minuscule worms an advantage when it comes to the catastrophes and mass extinctions that have struck Earth several times; it would be their refuge.

Borgonie was relieved his physically-punishing and time-consuming nematode bet had paid off. When future scientists start digging in earnest on Mars, he now had to believe, similar surprises could easily await them.

It turns out that extremophiles—mostly denizens of the bacteria and related archaea kingdoms—are everywhere on Earth. They have been found around scalding Pacific Ocean hydrothermal vents called "black smokers"—where "hyperthermophiles" live under pressure four hundred times greater than the Earth's surface atmosphere and temperatures at the vent mouths can reach 750 degrees Fahrenheit. Yellowstone National Park has also yielded organisms growing and reproducing in water hotter than 180 degrees Fahrenheit. The Rio Tinto in Spain flows red and is highly acidic (with a pH of about 2) yet it is home to not only extreme bacteria, but also to other highly unusual and more complex organisms. The uplands of the river have been mined for five thousand years, but the tailings created are not generally considered the cause of the extreme acidity; rather, it is the extremophiles in the river feeding on the iron and sulfide minerals in the riverbed that produce, and thrive in, the resulting acid bath. Researchers already know that microbes live in clouds and are sending balloons up to the stratosphere (one hundred thousand feet up) and finding living organisms there, too. Indian scientists launched a helium-filled balloon in 2009 and found microbes alive between 12 and 25 miles up into the stratosphere, where usually fatal ultraviolet radiation is strong. The implications for extraterrestrial life are pretty clear: Life, if given an environment even the slightest bit friendly, will find a way to adapt and survive.

In 1998, even before many of the most dramatic extremophile finds were made, a prominent University of Georgia microbiologist named William "Barny" Whitman asserted in the *Proceedings of the National Academy of Sciences* that half or more of the biomass on Earth, the intact cells of creatures, plants, and single-cell organisms, lives thirty feet or more below the surface of the land and four inches or more below the bottom of the oceans. Since then, experts have debated the details of the estimate by Whitman and his colleagues, but have accepted the conclusion that of the life on Earth—the cells that make up all the insects, trees, mice, birds, fish, bacteria, and us—about half consists of those invisible to the eye, single-cell organisms that live below the ocean and below the ground.

Their conclusion, so at odds with how most of us understand the planet on which we live, will perhaps make this additional insight into the extremophile world more palatable: Researchers are also finding microbial life in the ice of the world's thickest and coldest glaciers, in the so-called cryosphere.

The presence of organisms near the top of a glacier, where sunlight can warm and liquefy the ice, seems to make sense, as does microbial life along the glacier's ever-moving bottom, where the Earth's geothermal warmth and the glacier's friction create a somewhat watery environment. But recent discoveries have included the presence of organisms, or organism remains, at almost any depth measurable. While the research isn't conclusive, it certainly appears that some are not in a dormant or spore phase, but rather are actively metabolizing chemical impurities in the ice, using them as an energy source to maintain the disequilibrium needed for life and to repair their DNA, and to consequently live for impossibly long periods of time. This most extreme of ecosystems is apparently made possible by an unusual but universal property of glacial and sea ice: Those seemingly solid blocks actually contain a maze of minuscule veins that can include liquid water in ice to very low temperatures. This can happen because the physics of ice crystallization pushes all extraneous salts, minerals, and organisms into these microscopic pathways, creating environments where a microbe would have the nutrients and the unfrozen water needed to live. Some of the pioneering work in this field is being done in the unlikely and balmy setting of Baton Rouge, Louisiana.

On the day I visited Louisiana State University and its glacial life program, it was in the seventies outside. Students were mostly wearing shorts and T-shirts, and the world was green and soft. In the cold room where program director Brent Christner and his students do their work, the temperature was a steady 23 degrees Fahrenheit. We had on parkas and lots of polar fleece. Working in the icy cold has left at least one of the grad students perpetually chilled and wearing a down parka as he walks around campus.

The two-hundred-square-foot cold room is for studying the ice and houses a workstation with high-precision air quality control, a light box, a band saw, a subzero growth incubator, and an air vent—under which the "windchill factor" drops even further. But the real treasure trove is through another thick, tightly latched door into a –20-degrees storeroom filled with ice cores and ice blocks from around the world, and especially from Antarctica. Wrapped in heavy-duty plastic bags and stored in see-through container boxes from Home Depot, the ice looks as lifeless and exciting as cement. But Christner knows better; years of work have shown that they contain microorganisms barely eking out a living, yet apparently metabolizing, using energy, maintaining their DNA and splitting—all at a, well, glacial pace. Christner is confident this is correct because he has observed microbes that have been frozen in ice for many thousands of years start wiggling when warmed up a bit in the lab. Working with Mark Skidmore of Montana State University, Christner and his team, the Interdisciplinary Collaboration Investigating Biological Activity in a Subglacial Environment (ICIBASE), are finding recently unimaginable activity at the icy boundaries of life.

Christner has made five trips to Antarctica between 2000 and 2009 searching for life in ice. Scientists like Christner especially love the McMurdo Dry Valleys of Antarctica because of its many extremes: It's a two-thousand-square-mile region of snowless, gravelly valleys on a continent otherwise covered in white, it has liquid lakes hidden beneath deep coatings of ice, scores of glaciers, and winds of up to two hundred miles per hour. Not that long ago, the region was considered to be devoid of life. Now teams go down looking for, and finding, all kinds of microscopic organisms. It's also a mecca for scientists and engineers preparing for the day when Americans (or others) will need to know how to operate, and what to look for, in the environments likely to be found on Mars and elsewhere beyond Earth. Just a few miles from the ICIBASE camp, a NASA-sponsored team was continuing its long-term research with technology they hope will one day travel to Jupiter's moon Europa and be used to explore the enormous liquid ocean known to lie beneath its thick crust of ice and suspected to have conditions suitable for life.

Making the LSU research station anywhere near as frigid as where the ice came from requires heavy-duty condensers, compressors, and fans and results in a constant hissing sound—adding to an already surreal disconnect between inside the room and out. It's a small place that combines the feel of an operating theater and a meat locker, minus the hanging carcasses. Here ice is the star.

An ICIBASE alumnus hauled out three thirty-pound blocks from the storage room, each from a different part of McMurdo's Taylor Glacier, and placed them on a light box. One was clear, though full of captured, cartoonish bubbles; another had thin layers of sediment that produced an elegant, soft layered effect that could have come from a potter's hand. The last sample came from the bottom, where the glacier—a river of ice—grinds the rock below and mixes with it. Christner was most interested in the second sample, the one with the thin layers of sediment.

"We can melt the ice and bring the microbes out—they're alive, we know this," he said, gesturing to the block on the light box. "We can also take a natural ice sample from Taylor Glacier and measure the gas concentrations, and I can tell you in a piece of ice like this they make no sense at all. The level of CO_2 is [three thousand] times what it is in the atmosphere now, so something caused it to increase. That value couldn't possibly be atmospheric from earlier days because it's way too high. Oxygen is normally twenty percent of the total gas in ice, exactly like in the atmosphere. But in this area of our sample, the O_2 depleted. So you have CO_2 increasing and oxygen decreasing—it's a classic signature cellular respiration."

Okay, the microbes may be respiring (breathing, if you will) in the ice, but how could they possibly reproduce and keep their community from disappearing? That takes far more activity than these organisms could seemingly muster in their barely liquid, very salty ice vein habitats. Christner had an answer, one based on measurements of the rate at which microbes, living in ice at 5 degrees Fahrenheit, can build the genes needed to successfully divide and create a new organism. "There's a misunderstanding about microbes, that they're always dividing. But like humans, they're not reproducing all the time at all in nature. Okay, these guys are extreme—they may divide once in two hundred years. Just do the math: We know that some cold-adapted bacteria can synthesize about one hundred base pairs of DNA per day. But they have a genome with about three million base pairs, so it takes a while—not the kind of project a grad student would want to start up and ever expect to finish. Time means nothing to microbes. It's all about maintenance, just keeping alive."

When ice from the latest expedition arrives, the room will have more than 1,500 kilograms of ice that could contain something on the order of 15 billion microbes—all in some stage of "living," but nonetheless frozen in ice.

Pulling that precious ice from deep inside the Taylor Glacier was quite an operation, one that required two years of planning, a fair amount of equipment, and the help of National Science Foundation helicopters. It was frigid and windy when the team arrived at their site, nearby Lake Bonney, and set up camp. Each day they hiked a mile up to the face of the glacier and got to work. Using chain saws, demolition hammers, and ice picks, the group had to first build an ice stairway up about thirty feet from the base of the glacier. Then, over a week, they dug a tunnel more than forty feet into the ice—pulling out more than fifty tons via banana sleds. Above the tunnel was another hundred feet of ice, and picking the wrong spot to dig could lead to a catastrophic cave-in. That's why Christner and Skidmore helicoptered out from the American base at McMurdo Station and spent four hours intently surveying the glacier's ice face, looking for a spot without any telltale signs of surface weakness or calving.

The ICIBASE team—five men and two women—took turns with the equipment, each cutting and drilling until they were coated in ice chips and dust, and looking rather crazed. But the chain saw work was generally considered plum, because inside the glacier they were protected from the cutting winds that froze the haulers and lookouts on guard for teammates in distress, with windchill temperatures down to −40 Fahrenheit. At the end of their ice alley, the team studied the frozen walls for unusual and differing traits, and cut and drilled and yanked out slabs of up to one hundred pounds, for transport back to Louisiana and Montana for study. Two years before, they'd had to donkey-haul their catch out of the tunnel and through a boot-grabbing mudslide because the glacier had begun its yearly melt early, and a waist-high river had quickly formed between their ice castle and dry ground. This year they arrived sooner so they could finish the tunnel and airlift out the blocks before the melt river appeared.

The goal of the mission was to determine whether microbes in the ice constitute an archive of dead or metabolically inactive organisms, or if they formed a living, interacting community. Previous work in Antarctica had identified seemingly active biochemistry—the presence of by-product carbon dioxide and nitrous oxide with distinctively life-produced signatures, akin to the gases ICIBASE had found in the Taylor Glacier ice—and raised the prospect of finding microbes with clearly identifiable forms and functions. In other words, Christner and a handful of others are working to prove the hypothesis that Antarctic ice, as well as glacial ice elsewhere, is not lifeless and unchanging but rather an ecosystem no different from a forest or stream. It's clearly an extreme, slow-moving, and spare environment, but it's a world that supports certain kinds of similarly constructed life just the same—organisms with the kind of antifreeze found in some cold-water fish, organisms that appear to depend on the biochemistry performed by other nearby organisms for their survival, organisms with an ability to withstand extreme desiccation and intense radiation. That trait is essential, since scientists assume they initially blew down to Antarctica from the oceans or other continents, withstanding long, harsh periods in the atmosphere. The definitive research has not been done yet that would prove glacial microbes are conducting the work of "life" in the ice—as opposed to what they do when they're brought into the lab, fed, and warmed up a bit, and begin to move around—but the logic of the argument is getting stronger with each expedition.

And how extreme can their Antarctic living conditions get? Russian and American researchers, including Christner, have identified signs of life almost three miles below the surface in ice that originates from liquid Lake Vostok, the largest of the recently discovered subglacial lakes on the continent. That's nearing halfway up Mount Everest if you're going in the other direction. A high, windswept plateau of East Antarctica, the surface of Vostok is often described as the coldest place on Earth. The irony is that the deeper the drill goes into the ice above Lake Vostok, the more likely that they'll find living microbes. The reason is that it gets warmer deep down in the glacier, and finally warm enough under the ice that Lake Vostok (roughly the size of Lake Ontario) stays liquid all the time. The weight and pressure of the glacier clearly are part of the reason why, but Russian scientists—who have worked at Vostok since the mid-1950s—believe geothermal hot spots under the lake may be spitting out heat and gases as well.

Russian scientists and engineers are scheduled to make their long-delayed piercing of the pristine lake in 2011, bringing up what will be the first Vostok water ever touched by humans. Astrobiologists and Antarctica specialists are torn—eager to know what might be living in the waters, but worried that the Russian drill and collection device will contaminate the lake. While Vostok is the largest subglacial lake in Antarctica, there are hundreds of others. One of them, Lake Bonney in West Antarctica, has already been explored and mapped by a NASA-sponsored mission headed by Peter Doran of the University of Illinois in Chicago and underwater-robot maker Bill Stone, one of the breed of out-of-the-box engineers drawn to the same challenges and questions as Christner and other scientists. Without men like Stone to design, calibrate, and operate robots and machines that allow scientists to make finds in the most inhospitable parts of Earth, the most revelatory astrobiology expeditions would never have gotten off the ground.

Tall and lanky, Stone is a grown-up version of the science geek who had a better chemistry lab in his basement than his high school did. When he was still in school, his mom suggested he join a club to meet other kids. Not knowing what it was, he chose spelunking. Some decades later he is still a caver, leading extreme expeditions into the deepest caves on Earth: weeks of rappelling into the dark, swimming through narrow water tunnels, sleeping well below the surface on ledges and outcrops sometimes never before visited, and then climbing back up sheer cliffs. When not caving, Stone runs Stone Aerospace outside Austin, where the Lake Bonney robot was conceived and assembled.

Doran and Stone's Lake Bonney robot logged 250 hours under the ice and traveled more than thirty-five miles to conduct the first sophisticated, 3-D exploration of a polar lake, one covered in about ten feet of ice. The Volkswagen-sized "Bot" was also a test of concept: Could a submersible robot be programmed to not only follow a preset path, but also to independently explore areas where something unusual appeared? During two seasons under the ice the Bot, with the help of forty-five computers, performed as hoped.

But the real prize is hundreds of millions of miles away, on Jupiter's moon Europa. Beneath miles of ice, Europa is home to a giant ocean sixty miles deep and with twice as much water as the oceans of Earth—a prime target for astrobiologists. A joint NASA-ESA mission to Europa is now in the works for the 2020s, a scouting party preparing for the day when the moon's ice cover will be broken and a much smaller and more sophisticated bot than Lake Bonney's will be inserted into the waters below. A liquid ocean, even if it's kept dark by miles of ice, is the kind of place where life just might exist; in fact, we know it does exist right here on Earth.

Back in Baton Rouge, Christner might as well be in a different galaxy. He favors shorts and loud shirts—an azure Hawaiian shirt the day we met—and still puzzles about exactly how he became an ice man. His early great interest, what brought him into microbiology and then astrobiology, was thermophiles—the extreme organisms that live in and around hot springs, deep ocean vents, and the outer reaches of volcanoes. On schematics of the tree of life, many thermophiles can be found at the very bottom, leading some scientists to speculate they were the earliest or at least among the earliest life forms on Earth.

"Yes, I wanted to study thermophiles for my doctoral thesis," Christner said, and laughed. But instead he went into the microbes-in-ice field, in part because his mentor at Ohio State University had a collaboration with another team that had a large collection of ice cores from around the world. "They had ice from the Andes, from Tibet, Greenland, and Antarctica; and it was a huge opportunity. But to tell the truth, I thought when I was done I would write up the findings and move on to something else. My assumption was based on the conventional wisdom of the time that any microbes in there were just hanging out in the ice, doing nothing. We could learn about the past from them, but nobody thought about anything beyond that. Bottom line, that's not at all what we found. We found things alive in the oldest ice, seven hundred and fifty thousand years old. Some were spores, but some were not and seemed to be doing things, especially the ones near dust and sediment."

His tentative hypothesis had been that glacial ice is generally not a reservoir for dead or hibernating microbes, but is instead an environment with a complex and extensive living ecology of its own. Five trips to Antarctica later, he is increasingly confident that this is true. In fact, based on what he acknowledges is very limited data collected from two drilling holes, he believes Antarctica—which contains about 70 percent of the fresh water in the world—may well sustain a world of microbial life that exceeds in mass all the life found in all the freshwater rivers and lakes of the world combined. In samples from the Vostok ice core, almost three miles down, Christner isolated a bacterium that produces a protein that may well help the organism to survive, in part by altering the freezing and recrystallizing of the nearby ice—directing the potentially destructive processes away from the organism. In other words, their survival advantage appears to have come from a molecular adaptation that allowed the microbes to change their frozen environment enough to make life possible.

That kind of remarkable ability to inhabit such a harsh environment, and to even seemingly transform it, is why Christner assumes microbes will one day be found on Mars and elsewhere. "Based on what we're learning on Earth, I can't see that microbes living on Mars would be such a big jump. I mean, the conditions they live in here are in some ways just as severe, yet they've adapted. Not only that, we believe that ice is a habitat for life—providing a liquid environment under otherwise frozen conditions. Waste from one microbe is food for another, so it is likely that microbes are interacting and cooperating to extract every usable bit of energy. Almost all the water we see in the solar system is in an ice form, and I don't see why some of that ice wouldn't have microbial life, too."

It's that kind of leap of the imagination that has people like Christner looking deep inside the Earth to see life in the beyond. To enter, explore, and ultimately understand the world of microbial extremophiles, you need to be, inside your head, something of a human extremophile yourself.

3 WHAT MAKES SOMETHING ALIVE?

What is life? Even though an answer has been passed on to generations of biology students, they weren't getting the full story. When scientists invented the modern field of astrobiology, they had to wrestle with a fundamental problem: There is no scientific consensus about precisely what makes something alive. Given that unsettling absence, did it really make sense for astrobiologists to apply to the rest of the universe the never-quite-exact definitions we had come up with on Earth? To make matters all the more confounding, what would be a sure signature of biology on our planet could be totally nonbiological on Mars, and vice versa. So how do you find life in the beyond if you can't agree on what life is on Earth?

Because it's in the business of trying to find "life" beyond Earth, NASA has probably done more to try to define it than any other organization. Here is an unofficial working definition: Life is "a self-sustaining chemical system with the capacity to evolve in a Darwinian manner." The definition came out of a workshop of biologists, physicists, and chemists in 1994, and it does meet many of the basic criteria scientists and others are looking for. Broadly, it accounts for the known constants of life on Earth. All living organisms take in some form of energy, use and change it, and then release it as waste; all use the same twenty amino acids to construct the proteins that make that and all other activity possible; and all use RNA and DNA molecules to store genetic information and to construct proteins. The Darwinian evolution comes directly and inevitably from the presence of DNA, since all DNA mutates.

But the definition has many critics, some of whom think it is not only incorrect but also misguided. The criticisms come from many directions: those who argue the definition would rule out viruses, prions (which cause "mad cow" disease), and other seemingly "living" organisms; those who want to base any definition on a specific capability such as metabolism or reproduction or the enclosing of a cell nucleus by a cell wall; those who think in the more abstract terms of a physicist and want a definition that takes their discipline (the Second Law of Thermodynamics, for one) into account. Relying on that law, the Austrian quantum physicist Erwin Schrödinger famously suggested that life—in its broadest terms—be defined as something that avoids immediate decay into "entropy," the chaotic and then utterly uniform state the entire universe will someday revert to since all structure has in it the seeds of its own falling apart. Living things, Schrödinger proposed in his 1944 book, *What is Life?*, postpone this inevitable process by taking in nutrients and turning them into energy; at death the life forms eventually succumb to the force of entropy and break down so the atoms of the once-living body become evenly distributed again, recycled by the Earth.

Portland State University geobiologist Radu Popa, author of the 2004 book *Between Necessity and Probability: Searching for the Definition and Origin of Life,* said that he lost count of the proposed answers in the scientific literature after logging at least three hundred. And the definitions keep coming. Nilton Renno, a planetary and atmospheric scientist at the University of Michigan and a member of the Mars Phoenix lander science team, recently came up with this one in a paper on the likelihood that the heat from the spacecraft's landing created liquid water that remained visible for days: Life, he wrote, is a self-replicating heat engine with a capacity for mutation.

Perhaps the most subversive challenge to the proposed definitions of life comes not from those who think the NASA definition is incorrect, but rather from those who think "life" is not a concept we can or should define. Philosophy professor Carol Cleland, from the University of Colorado, and Chris Chyba, an astronomy student of Carl Sagan's who now teaches at Princeton University, have argued for almost a decade that current definitions of "life" are little different from medieval definitions of "water," which was seen then as a clear liquid with certain qualities such as wetness, transparency, tastelessness, odorlessness, and the property of being a very good solvent. We can now chuckle at the misunderstanding, since muddy water is certainly not transparent, salty water has a taste, and marshy water has a smell. Medieval alchemists classified nitric acid and some mixtures of hydrochloric acid as *aqua fortis* (strong water) and *aqua regia* (royal water) because they were such good solvents.

But water, as we now know, is H_2O—two hydrogen atoms bound to one oxygen atom. Those men and women trying over the centuries to define water knew nothing about the molecules and atoms that we now know make up all matter. That didn't come until the late eighteenth century, when Antoine Lavoisier came up with the convincing theory that matter is made up of molecules. Cleland, Chyba, and others have argued that the basic knowledge needed to make a definition of "life" is simply absent, rather like how the essential molecular nature of water was unknown during the Middle Ages. Based on her iconoclastic views—grounded in philosophy and at times a challenge to scientists—Cleland was included on a University of Colorado astrobiology team that was twice funded by NASA. Her thinking became more broadly known when she addressed a 2001 meeting called "The Nature of Life," hosted by the American Association for the Advancement of Science. She told the audience of scientists that the search for a definition of life—something many were involved in—was a waste of time and, even worse, misleading.

"The logic of my argument was impeccable, but people just blew up at me," recalls Cleland, an expert in the philosophy of science. It was a memorable evening. "They were yelling out their own definitions, saying this is the right definition or that is the right definition. It's as if they totally missed my point that their approach was mistaken and there is no definition available now. I was kind of shocked and remember saying to myself, 'These people just can't hear what I'm saying.' I've learned since then how to better talk with scientists, but I still think the whole definition project is hopeless."

Ten years later, Cleland and Chyba's view is no longer outlandish. Addressing a NASA-NSF gathering of many of the nation's top practitioners of "synthetic biology" (the origins-of-life side of biotechnology), the evolutionary biologist Andrew Ellington, of the University of Texas at Austin, urged NASA to bring together a blue-ribbon panel to study and then throw out the agency's and all other definitions of life.

"It is my position that there is no such thing as life, and that the working statement in the NASA document does science a disservice by attempting to pretend the contrary," he told the gathering of in 2008. " 'Life' is a term better suited for poets (or perhaps philosophers) than scientists, and the continuing attempts to determine whether a given system is alive or not harken back to quite ancient philosophers, with a similar level of resolution. I assert the following existence proof: if we haven't figured out what life is by now, there is little hope that we will figure out a definitive definition in the near term, and there is no research program that I can imagine, at any price, that will provide such a definition."

Ellington then made clear why he felt as strongly as he did. As is so often the case in astrobiology, the purely scientific issues are surrounded by deeply felt and highly contentious social and even political issues. "I would further argue that the reason that what is nominally a rather pointless philosophical issue has become an important one for NASA is because of its near-term political ramifications," he said. He believes defining "life" is a dangerous endeavor because the information collected will almost inevitably weigh down science. "I can imagine a day when the head of NASA would be brought before the Supreme Court in an abortion case and asked to define life," he told me. "And I can imagine the long and uncomfortable silence that would follow." Let the work progress on synthesizing molecules that can do what living molecules do, and on determining if some unexpected substances have lifelike qualities, he says. But leave the definitions for later.

The controversy over a definition for "life" has actually been around for some time, even inside NASA, and it became a serious problem and even embarrassment in 1976 when the agency landed two Viking spacecrafts on Mars in a self-described search for life. To the initial delight of the Viking scientists, a key biology experiment on both Viking landers gave a strong signal that "life" had been found—meeting the painstakingly crafted criteria established before the spacecraft left Earth—and the control experiments seemed to confirm the finding. Yet the principal investigator of that experiment was held back from announcing what Viking had apparently discovered. The scientific community and NASA quickly formed a consensus that life had not been detected. The problem wasn't with the way the instruments performed or how the experiment was carried out, but rather with the definition of life that NASA itself had put together, one based on the way metabolism is known to work on Earth.

The story is best told through the life and times of Gilbert V. Levin, a pioneer of astrobiology who began his career as a sanitary engineer searching for microbes in drinking water. He first proposed a life-detection experiment for Mars in 1959 and had his idea embraced and tested time and again by NASA before the Viking launch in 1975. He got the results he had dreamed of within ten days of the first landing of Viking. It seemed like a scientific triumph of historic proportions, but it quickly slipped away and Levin has been fighting ever since to reclaim the victory. More than ever, he says, he is convinced that his Viking experiment did find something that indeed was—had to be—living. But the scientific verdict came down against him and, despite some converts, has not significantly changed.

Levin's experiment was conceptually quite simple: It added a number of liquid nutrients that had been "labeled" with radioactive carbon 14 to a sample of Martian soil dug up by the Viking collecting arm and pulled into the spacecraft. If these nutrients were eaten by Martian bacteria or other life forms, the gases they would inevitably release as waste would also be radioactively labeled and would be detected by an installed radiation counter. It was a simple and powerful test for a cornerstone of all definitions of "life"—the ability of an organism to use the chemicals contained in food to produce the energy it needs to maintain itself, to grow, and to reproduce. If radioactive gases were released, Levin and his NASA collaborators initially agreed, then an organism had taken in and broken down the nutrient food, and was passing the waste out when it was done. Thus the experiment's name: Labeled Release.

After the nutrient was squirted into the soil collected on Mars, the monitoring instruments registered a surging amount of radioactive carbon dioxide gas—strongly suggesting that some organism had eaten the food and then released the gas. A follow-up control experiment heated the soil to a high temperature that would presumably kill any living organisms, and then squirted in the nutrient. This time there was no release of CO_2, an apparent confirmation that the gas had been produced by the actions of an organism that had been alive during the first experiment but was killed by the heat in the second. Viking 2 landed four thousand miles away on Mars a month-and-a-half later, and the same Labeled Release experiment was conducted. Again, the radioactive gas was detected when food was delivered to the Martian soil at what amounted to room temperature, but not after samples of the same batch of soil were heated and cooked, or when it had been stored in a dark container for several months. It certainly seemed that metabolism—a process only known to occur in living organisms—was taking place. Two other Viking biology experiments got strong reactions when food was presented in gaseous form to the soil, but the controlled versions failed to support the results. Scientists quickly concluded the reactions came from chemical, and not biological, sources. Nonetheless, Levin was convinced that he had found life on Mars.

NASA was skittish about Levin's results from the start. Officials cautioned that all the reactions could be chemical rather than biological, and that the speed of the appearance of radioactive CO_2 did not appear consistent with a biological reaction (although they admitted it wasn't consistent with a known chemical reaction, either). What they needed to make a firm scientific judgment was the data coming from another key experiment, one designed to determine whether organic compounds—the carbon-, hydrogen-, and oxygen-based molecules essential to all life on earth—were present in the soil. That experiment used a gas chromatograph mass spectrometer (GCMS) to heat the soil until chemicals turned to vapor, and then it separated, identified, and quantified the large number of different chemicals found. The device, refined and operated by prominent MIT biochemist Klaus Biemann, was designed to measure molecules present at a level of only a few parts per billion. The Viking arm twice failed to bring in soil for the GCMS, and so NASA and the many Viking watchers had to wait for days before the testing could begin.

When the samples did arrive, the results were both surprising and seemingly unequivocal: The instrument measured no indigenous organic molecules in the soil, indicating that Martian soil had even less carbon in it than the barren lunar soils brought to Earth during the Apollo program. The strongest organic concentrations it measured were minute trace chlorine-based organics written off as contaminants brought from Earth. Without organics, the scientists concluded, there could not be life, and so any experiment suggesting otherwise had to be reinterpreted. The anticipation that Viking just might delight the world by finding life on Mars quickly turned to a conviction that Mars was lifeless—without organics, without water, and seemingly with compounds all around that rapidly bound other elements to oxygen and made them inaccessible to potential life. A consensus quickly formed that the reactions in Levin's experiment and the others had to be chemical and not biological, and that's the way the Viking results were presented to the world and understood by the scientists—all except for Levin and a handful of others, that is.

• • •

The fact that at both Viking sites radioactive carbon dioxide appeared in significant amounts during his experiment and didn't appear during the controls, that the experiments met all the criteria set out before launch for a positive finding of biological activity and life, was too much for Levin to leave undefended and let go. And so for more than thirty years he has done just that—reminding one and all about the Labeled Release results, citing tests of his experiment in extreme environments around the world, and working hard to knock down all the alternate explanations offered. Others have joined the fray in recent years, and Biemann's mass spectrometer has been found wanting in a number of reviews by respected scientists. Those men and women don't necessarily endorse Levin and his conclusions, but their research found numerous instances where the GCMS instrument would (in theory) and did (during testing) miss the presence of certain organic compounds in extreme Earth environments, especially when their concentrations were low. A 2010 paper by two prominent astrobiologists, Rafael Navarro-González of the National Autonomous University of Mexico and Chris McKay of NASA's Ames Research Center, went further: They concluded the GCMS actually destroyed organics by heating them. And the chlorine-based organics that Viking scientists wrote off as trace contaminants from Earth were precisely what would be left behind if Martian organic material were heated along with surrounding Martian soil.

Even Biemann, who defends his Mars work vigorously as having determined that the Viking landing sites could not and did not support life, nonetheless does not believe it represents a final word on Martian biology. He ended a recent defense by writing: "Future missions to Mars will sooner or later answer the question of organic matter at the surface or in the near subsurface of that planet. It will require carefully designed instrumentation to carry out well planned experiments and thoughtful interpretation of the resulting data." The implication, it would certainly seem, was that Viking did not meet that grade. The next NASA mission to search for Martian organics, the Mars Science Laboratory, will launch in 2011 and has a similar if more highly evolved GCMS that can test for organics (and unofficially for signs of life) using solvents rather than heat.

Levin, born in 1924, is now an adjunct professor at Arizona State University and has long run a firm based outside Washington, D.C., that discovered and developed a low-calorie sugar called tagatose now in final clinical trial as a diabetes drug. Behind his gentlemanly demeanor, he is a scientific warrior. When the principal investigator of the 2008 Phoenix mission to Mars told a TV interviewer that the lander was the first to touch frozen water on Mars but that the planet has no liquid water, Levin had a rejoinder on the show's website within six minutes. "What a comedy!" he wrote. "Liquid water was discovered on Mars by the Viking lander in 1976! Ice was shown in images taken by the lander. We have published several papers proving liquid water on Mars. AND we claim that our Viking Labeled Release experiment detected living microorganisms on Mars. . . . Paradigm shifts are difficult, but this one has taken way too long!"

Levin had just moved into a modest town house outside Washington when we first met, and many of his various scientific trophies and memorabilia were still in boxes. Although he has three degrees, including a doctorate from Johns Hopkins University, he believes that his beginnings in the world of sanitary engineering and that his Mars research was not done at a prestigious university are held against him.

"I've thought long and hard about this and I think that when the Viking results came in, NASA was confronted with evidence of life and no evidence of organics. One result came from a prominent professor and another from an unknown guy from a small company. The more conservative folks were more comfortable with Biemann and his 'no organics,' and that was the ballgame." Nobody seriously questioned Biemann's instrument until years later, when it was shown by several teams that the instrument could not detect very low levels of organic material in samples from Earth known by other means to have living microbes in them. By then decades had passed, and NASA was stuck with its no-life position because overturning it would raise a new set of other controversies. Levin's conclusion: "Nobody in charge was brave enough to say it was wrong and NASA still doesn't want to go near the issue. After Viking and until the present day, there have not been any life-detection experiments sent to Mars, even though finding life there would be the biggest discovery in the history of science."

Not surprisingly, many see the Viking results and subsequent scientific approaches to Mars quite differently. For instance, Michael Meyer, the lead scientist for NASA's Mars program, and who has a longtime involvement with astrobiology, said an essential lesson of the Viking missions was that we don't really know how to look for life yet, an embrace, of sorts, of the Cleland and Ellington position. Levin's experiment focused on an undisputed signature of life—metabolism—but Meyer says the results were positive but ultimately not convincing. "He might have found life," Meyer said, "or he might have found that nonbiological processes take place on Mars very differently than they do on Earth." The release of CO_2 could have been the result of a not-yet-understood chemical reaction, for instance, if compounds with a lot of free oxygen were present. In other words, what would be a clear indication of biology and metabolism on Earth could be totally nonbiological on Mars. The upshot of the Viking life-on-Mars debate has been that NASA has studiously avoided sending life-detection experiments to the planet ever since, choosing instead to concentrate on geology, mineralogy, weather, and the search for water present and past.

In the early 2000s, the United Kingdom sent Beagle 2, a small probe designed to look for life-sustaining habitats, to Mars, and Levin tried without success to get a life-detection instrument into the mix there as well. Speaking with BBC News before the planned landing, deputy mission manager Mark Adler explained that Beagle's mission was to better understand the water environment of Mars and not to search for life as Levin urged. "What we learnt from Viking is that it is very difficult to come up with specific experiments to look for something when you don't really know what to look for." But it all became moot when Beagle's mission control lost contact with the spacecraft as it entered the Martian atmosphere, and disappeared. Levin did succeed in getting a bare-bones life-detection experiment onto a Russian mission to Mars in 1996, but that effort failed before it even reached the planet.

Still Levin is seeking vindication and has (among others) his physicist son Ron Levin working with him. In 1986, the senior Levin told a Viking ten-year reunion gathering at the National Academy of Sciences that "it is more likely than not that the Viking LR detected life." In 1997, he argued in a paper for the *Proceedings of the International Society for Optical Engineering,* which society has an active astrobiology program, that twenty years of additional Mars research had convinced him that his Viking experiment *had* definitely detected life and that NASA and the scientific consensus were wrong. Nine years after that publication, with an appreciation of Levin's work emanating from a new generation of Mars scientists, an Argentinian scientist proposed the name *Gillevinia straata* as the genus and species of the bacteria-like organism ostensibly identified by Viking. But that idea did not garner much support. Levin was not invited to give a talk at the official thirtieth anniversary of the Viking mission.

To this day, Levin is not inclined to think that the absence of a firm definition of "life" played a significant role in the scientific community's reluctance to accept his Viking data; it's something of a red herring, he says, used to protect important people from having to admit they were wrong. But that refusal to entertain other possibilities, to essentially reject the notion that testing for life on Mars might require a different way of thinking, is what frustrates many scientists about Levin. Yes, Levin is tirelessly and heroically defending a result defined at the time as positive. Yes, the experiment uncovered something of great interest, and nobody has been able to explain why the Labeled Release and its control behaved as they did.

But thirty years of additional research and thinking about Mars has, in many ways, turned Viking's simple models about life on their heads. As Meyer explained it, "In some ways, you could say that Viking was too Earth-centric. It presumed life has metabolism and respiration that results in production of carbon dioxide that we can recognize. It also presumed that if you land anywhere on Mars you can measure life." He said that while NASA has at times used the definition of life as a "self-sustained chemical system capable of undergoing Darwinian evolution," it was not a formal position, and the agency was increasingly inclined to accept the reasoning of Cleland and others that "life" cannot be currently defined, any more than water could be in the sixteenth century. "Probably the characterization people are most comfortable with is the Supreme Court one on pornography, that 'we know it when we see it.' But for a variety of pretty obvious reasons, that one really didn't fly." Describing essential characteristics of life—that's certainly possible. But a final, all-encompassing definition that provides the invaluable solid ground scientists have been searching for, that will have to wait.

As practitioners of another arm of astrobiology will quickly point out, you don't have to go to Mars to get confused about whether something is alive—that is, the result of biological processes—or not. A parallel and sometimes equally intense debate has been going on for several decades about a substance found on Earth called desert varnish. Best known as the purple-black background to many ancient American Indian drawings (or petroglyphs) in the southwestern United States, the varnish coats rocks in arid climates. Nobody really knows how it gets there. Researchers have found large colonies of bacteria living beneath the very thin yet definitely layered varnish, and they have found very high concentrations of the element manganese in the coverings as well.

Living things have the property of concentrating elements in ways that nonbiological processes do not, and so the unusually high levels of manganese and sometimes iron in varnish—much higher levels than in the surrounding environment—definitely suggest biology. The opposing line of thought is that the bacteria found under the varnish have come from elsewhere and have simply found a protected place to live in very harsh environments. As for the high concentrations of those elements not concentrated in the soil or other rocks nearby, those are the result of chemical reactions and collections of windblown dust. You would think this would be a relatively simple question to answer, but it isn't. Desert varnish thickens at an extremely slow pace, on the order of between 1 and 40 micrometers (or 0.000003937 to 0.00015748 inches) per every thousand years, so it has been impossible to experiment definitively with it in the lab. After centuries of growth, a rock's varnish covering will be about the thickness of a piece of paper. And nobody knows how or why it spreads.

As is often the case in astrobiology, the players in the desert varnish story come from a broad range of backgrounds—specialists in caves, in planetary science, in geography, in engineering. They get pulled in, not only by a desire to unravel the mystery of the origins and nature of desert varnish, but also because of some images of rocks that have come in over the years that seem to show something that looks surprisingly like desert varnish in an unexpected location: Mars. Both NASA and the National Science Foundation have funded research into desert varnish, and some years ago it was a very hot topic. The combination of painfully slow progress in understanding how the varnish grows, along with a mini-scandal in the field regarding some questionable data, has pushed it to a back burner. But that doesn't mean some intrepid souls are not still surveying the murky borderland between biological and nonbiological life.

One is Penelope Boston of the New Mexico Institute of Mining and Technology in Socorro. She is an expert in caves and the microbes that live in them, but also has a passion about both Mars and desert varnish. A woman of many enthusiasms—her department office is overflowing with alien action figures, stuffed animal bats, robots, and name tags from hundreds of conferences around the world on caving, Mars, and astrobiology—she is happiest out as a field researcher. It was while doing a five-day research expedition in the Lechuguilla limestone cave in the Carlsbad Caverns National Park, the deepest cave in the nation, that desert varnish came into her life. She was quite deep in the cave when some "fluffy" greenish-reddish-purplish material swirling in the air fell into her eye. It didn't take long for her eye to swell shut, leading to a harrowing rope climb up and out of the cave but—more important to her—also yielding one of those "aha!" moments when it becomes clear things are not what they seem.

A trained microbiologist, Boston immediately understood that some microbes had gotten into her eye, which meant that they were living deep below the Earth's surface on what appeared to be rock face. This was before Onstott's South Africa work, so the scientific consensus was that nothing was alive in a deep cave, especially one known to be virtually locked off from the surface. Once she got outside, the swelling disappeared within four hours because, she also surmised, the microbe could not survive in the light of day. She and her colleague Chris McKay, of NASA's Ames Research Center, returned and over several years concluded that the fluffies were coming from the manganese and iron deposits in the caves, and that they were all part of a living microbial world. Driving around New Mexico, Boston constantly passed rock varnishes that featured the same manganese and iron that produced whatever had landed in her eye, and she got to thinking about what microbes might also be living in the varnish, or perhaps forming the varnish.

This encounter led to more than ten years of collecting samples of desert varnish and culturing them in her lab. The result is now a room filled with hundreds of stacked petri dishes, cylinders, and plates—some being warmed in incubators and some refrigerated—alive with what she is convinced are the bacteria responsible to a greater or lesser degree for the presence of desert varnish. The bacteria come from her home state, from Mexico, from Utah, from Chile, from volcanoes, from extreme environments of all kinds, and all are now growing in some agar medium and usually producing in bountiful quantities the purple-black signature of manganese. All started in a clear medium with tiny scrapings of bacteria added from a desert varnish sample, and most had produced massive (on desert varnish scales) amounts of the manganese-concentrating bacteria over the years. Surrounded by so much life, however microbial it might be, Boston talks to the samples, refers to them as "these guys" or "those guys," and says "we've got everyone in here." She knows which are "going gang-busters" and which are struggling to survive. One she describes as resembling, under heavy magnification, a bundle of grapes. "Look at this one," she says, pointing to a seemingly dead, crusted collection in a vial. "They look dead, but they're accustomed to desert conditions so they adapt. Rehydrate them a bit and they'll be growing fast, too."

Has Boston proven the desert varnish is indeed a product of living organisms? Not really. As she readily acknowledges, growing the samples demonstrates that the bacteria can and do concentrate manganese as contained in desert varnish, but that lab process has limitations. To achieve a level of proof, she would have to place her samples on rocks and see how they grow, and there's the rub: They grow at such a painfully slow rate that no professor, no graduate students would still be around to detect their progress. In Death Valley, it takes varnish something like ten thousand years to grow to a thickness of one-hundredth of an inch. She has yet to settle on a plan, but she has elaborate schemes for trying to force that growth in field conditions. That's probably to be expected from a woman who once lived two weeks in a simulated Martian environment in Utah and was one of the founders in the early 1980s of what became known as the Mars Underground at the University of Colorado, Boulder. A group of students who were fascinated by the planet and wanted, in the wake of the Viking disappointments, to keep interest in it alive, they sponsored a series of conferences that attracted prominent scientists and some NASA officials. Boston dreams of being around when life is discovered there; actually, she said she would gladly fly on the maiden yearlong voyage to Mars. In the meantime, as she tries to unravel through field and lab work the barest-bones life on Earth, she is getting help from another scientific outlier.

Tom Nickles has also been lured into the borderlands of desert varnish, and he is now conducting the experiment he believes will finally determine whether living bacteria or nonliving chemicals are responsible for the coverings. In a world of unusually bright people with unusual backgrounds, Nickles perhaps takes the cake. Tall and thin, he has wavy hair that served him well during his days as an occasional Elvis impersonator. He wears a belt with a large buffalo head buckle, which fits both his name and his locale—the University of Idaho in Moscow, Idaho. A trained engineer, former air force intelligence analyst in Turkey, test pilot trainer at Edwards Air Force Base in his beloved Mojave Desert (where he ran marathons), and so much more, he went back to school at age fifty to get a doctorate in astrobiology. There was no astrobiology program at the University of Idaho, but he found a professor who would sponsor him and he's now several years into the program. But he's hardly your typical doctoral student: He gets called in to do consulting for NASA and his plans for the desert varnish experiment looked so promising that he was invited to speak to the American Chemical Society's astrobiology panel before he even began the work.

In a stainless steel glove box with Plexiglas linings, floors, and dividers the size of a blanket chest, he had created an environment similar to the Mojave—with fans to simulate the wind, special lights to simulate the desert glare, and some extra moisture to speed up the growth process and simulate six years of varnish development in one. He divided the box in half, but both halves had a bed of sandy soil basalt and quartz as anchor for the varnish, should it grow. The two were identical except that on one side he planned to introduce some bacteria he had collected from varnish found outside Baker, California, at the Lima Lava Flow. The other side would have none of those bacteria. Would either, or both, lay down a varnish?

"Both sides start with absolutely nothing alive. The chambers have been sterilized and the substrate—the quartz and agate and basalt—have been autoclaved, so I'm sure everything will be dead. I will seal the abiotic side tight so absolutely nothing gets in, but on the other side I'll paint some of the bacteria I collected from varnish onto the rocks. Then I wait and watch. The fans will blow the dirt and dust around and the UV light will shine and the moisture will seep in and it will be just like the desert, except speeded up a bit."

The experiment hadn't yet begun when I visited; the varnished (biotic) samples were placed in the box a few months later. Nickles didn't expect to see black rock varnish anytime soon; that takes way too long. But the bacteria, with their extra UV light and moisture, could begin the unseen varnish-making process in months, or maybe a year. That process involves one of the most important dynamics of both astrobiology and geology: Life forms interact with minerals and rocks, and transform them in minute but detectable ways. They impose a distinctly biological structure onto their nonliving surroundings, and they also can (and usually will) concentrate certain elements or minerals in the process. In the case of desert varnish, the coating is blackened by a concentrating of manganese, or can come out reddish brown if the bacteria is concentrating iron.

"The experiment will go for a year, but I'll first open up the biotic side at three months. I'll take out three specimens at random and then will put them in the antechamber," an airtight but accessible cylinder outside the glove box. "After that comes the electron microscope to see if anything is being moved around. I'm looking for just a very crude laying down of structure, of organization, and the start of some layering. If biology plays a role in making varnish, we should start seeing [very early signs of activity] on the rock. Nothing beautiful like nice, glossy varnish forming. But something with structure."

Or maybe not. Maybe a varnishlike coating will emerge on the side without varnish bacteria instead, giving support to the theory that the varnish is formed by chemical processes involving the wind, the sun, dust, and the surface of the rock. And by implication, any varnishlike formations on Mars would have the same nonbiological origins. All of this raises the same fundamental question: If we can't determine or define what is life on Earth, how can we possibly do it on Mars or Europa or anywhere else?

On Earth, we at least know the basic molecular, chemical, and thermodynamic outlines of carbon-based life, but we are still in something of a quandary when it comes to definitively nailing it down. That's why scientists generally focus now on the known effects of living things on rocks, water, and atmospheres. We may not know what life is, but we have a pretty good idea of what life does—or at least the kind of carbon-based life found on Earth. What if extraterrestrial life is silicon based or has a very different way of holding and transmitting the information that produces future generations? The National Academy of Sciences formed a panel to study what came to be known as "weird life." It met periodically from 2002 to 2005 and released a report in 2007. Not surprisingly, it complicated rather than clarified the question of what life is by offering possibilities based on silicon (instead of carbon) and replacing water as the key solvent with ammonia or methane. Science has no examples of such life, but silicon has bonding properties similar to but less adaptable than those of carbon, and life based on a solvent other than water is also considered theoretically possible by some.

In science, the most desirable and convincing proof of a finding or theory is to replicate the reaction reliably under controlled settings. So some scientists are trying to make life out of nonliving elements and compounds in their labs. It's not exactly Dr. Frankenstein redux, but close. If they can create from component parts an entity that replicates, that takes in and uses energy, and that is able to both mutate and repeat that mutation with its replicator, then they can lay claim to having achieved a proof of concept. The creation wouldn't tell us how life actually began, but it would represent a process through which life *could* have begun. And along the way, it will help define, or at least to better characterize, what life actually entails.

It's a definite competition among some twenty of the world's most innovative and admired labs, but because the task is so daunting, it is also collaborative. But none of the scientists involved is as dazzling or as excitedly eclectic in their work as Steven Benner, a chemist and molecular biologist who created and runs the nonprofit Foundation for Applied Molecular Evolution in Gainesville, Florida. He keeps afloat on grants from NASA and the NSF, but also and most importantly on the profits from the creation of the world's first synthetic genetic system capable of producing unnatural nucleotides (the parts of DNA and RNA involved in pairing) used to monitor the levels of viruses that range from HIV to hepatitis B. With that support, he is able to push forward with efforts to produce that "self-replicating chemical system capable of Darwinian evolution" that defines life in its most broadly accepted form. With his wide-ranging knowledge and willingness to look seriously at problems from new, untried angles, he's often called on to help NASA and the National Academies of Science tackle big, complex issues; most recently, Benner was one of a handful of scientists asked by the National Academy to study the possible biochemistry of extraterrestrial life, the effort that produced the "weird life" report.

Benner's Gainesville lair is hardly the highly organized, precision-driven lab you might expect. Certainly the area where his almost twenty-person team of chemists and biochemists do their molecular slicing and dicing is well controlled, and the work of experts in paleogenomics (who read the evolutionary history of life-forms through their genomes) requires mind-numbing precision. But the heart of the operation, where the sparks fly, is Benner's office. At its center is a fifty-two-inch Sharp Aquos computer screen connected to a smaller one—the blackboard on which he diagrams chemical systems. He goes at the task with the focus and energy of an artist captured by a moment of creativity, and the screen can soon fill up with hundreds of connected *C*'s (carbon) and *H*'s (hydrogen) and *P*'s (phosphorus) as they interact and loop around to form known or possibly synthesized biochemical cycles. Benner, talking and writing nonstop, sits in an oversize pinkish reclining chair, taped in the back where the material is cracked. On prominent display around him is a sampling of his collection of minerals and fossils of fish and plants, ferns and small mammals. Benner has been a fan of both rocks and fossils since he was a boy, and each object has a story.

The boron-based rock on his shelf, for instance, captures one of his scientific eureka moments. Benner had been on Catalina Island, off Los Angeles, with a group of geologists, and he was leading them through some experiments involving the sugar ribose, a mineral with calcium in it, and water. The goal was to find a way to keep the combination from turning into brown tar, which seldom has anything useful in it from a biological perspective. This is a significant issue in the origins-of-life world because ribose is the R of RNA, and it has to be stable enough at some point to bond with the other elements. (Stanley Miller, of the Scripps Research Institute, outside San Diego—the deceased godfather of origins of life experimentation—famously found some precursors to the building blocks of life in 1954, opening the door to what was assumed to be a fairly imminent test-tube creation of life from nonlife. But it never happened and in 1995 Miller basically said it couldn't—primarily because ribose is unstable in the water that is assumed necessary to support life.) Benner long knew that the element boron at least temporarily blocked the decomposition of ribose in water, but he had never before thought to throw it into the origin-of-life chemical mix. But he did in Catalina, and the ribose did not immediately start the usual quick slide into tar when a pinch of boron was added. Instead it stayed clear, and a very excited Benner believed he had found a way to allow the essential ribose to be created on early Earth while still keeping water in the picture. This epiphany didn't solve the question of how life formed from nonlife, but it offered a plausible explanation for how one of many obstacles may have been overcome. When I first heard Benner speak at an astrobiology conference, he made a quick but quite serious aside suggesting that boron really could be central to the creation of life—even though it is one of the less common elements on Earth and across the universe.

The origins-of-life world has two competing schools. One says that the Last Universal Common Ancestor (or LUCA) formed as "life" when genetic material came together and self-replication could begin. The other view is that metabolism—the process of taking in the energy of food, using it, and then expelling a waste product—was the essential first step. Benner is in the genetics camp, which is why the origins-of-life component of his lab spends its time searching for ways to form the scaffolding of RNA or DNA out of nonliving parts. So many confounding factors are involved that biochemists like Benner are among the most skeptical about finding life beyond Earth. From the perspective of physicists, astronomers, biologists, and others in the astrobiology world, extraterrestrial life is a given. But to the chemists and biochemists working to actually get life started, the prospects are not as tangible. Making life from nonlife has turned out to be extraordinarily hard.

Benner says he assumes the actual origin of life—the pathway that created the first organisms that could feed, could replicate themselves, and then could evolve by producing mutant replications and keeping those that turned out to be useful—will remain unknowable. But finding another way that nonlife turned into life would be an entirely acceptable, actually enormous historical triumph, because it would provide the indispensable "proof of principle" that cutting-edge scientists are always looking for.

"Look, we know that life evolved from nonliving sources because otherwise we're left with divine intervention—which is hardly an acceptable explanation for most scientists. That's what keeps some of us going. It will be enormously difficult to find how it happened, but we know it *did* happen, either on Earth or elsewhere and then transplanted here."

The field has seen progress, even if it hasn't had the big breakthrough that many are looking for. In 2009, the journal *Science* published the results of work in the Scripps lab of Gerald Joyce, who with Tracey Lincoln produced an RNA enzyme (a protein that increases the rate of chemical reactions) that was a super replicator capable of building copies of itself over and over again, something never done before with RNA. This high-powered enzyme, refined and concentrated through cultures, met the primary goal of being able to perpetually replicate itself, as well as to mutate and then pass on the genetic blueprint of that mutation to other RNA. "This is the only case outside biology where molecular information is being passed through the generations [and] has become immortal," is how Joyce put it.

Joyce is now working to expand the functions of that immortal replicator and to see if it could survive and prosper in a more varied and complex system, if the stronger enzymes created by the lab could be challenged to invent new functions, just as RNA material presumably had to do on the early Earth. If he succeeds, he said, the lab will have indeed created life. "We believe genetic material that can respond to increasingly complex challenges represents life." I later asked Benner if that would meet his criteria, and he gave a quick "yes" as well.

But synthetic biology has an Achilles heel, hidden in plain sight: All of the researchers in the field make their creations using strands of preformed DNA and RNA that they can buy from a supply house. Manipulating those strands to make something out of them that can copy itself and somehow get energy from its environment would be an enormous achievement, but it would still require the scaffolding of genetic material to be delivered by overnight mail in a vial. Clearly, that's not how life started. Producing "life" in a lab may soon turn out to be possible, but it would require not only a lot of already formed complex molecules but also a lab full of scientists and equipment. Synthetic biology is science at its most imaginative and sophisticated, but at bottom it can't exist without the kind of intelligent designer that would give comfort to those who subscribe to a religious view of the origins of life—it requires a Creator. So the goal is a proof of concept, not a re-creation of the origins of life. Matt Carrigan, an origins-of-life researcher in Benner's lab, talks of research that would "jump the chasm," that would put together unprocessed molecules—not from the supply store—in a way that would allow them to begin life. But that kind of biochemistry seems very far in the future.

Ironically, it is the search for life beyond Earth, rather than the quest to synthesize it in a lab, that may hold the greatest promise for determining what constitutes life. On Earth, we have one essential model of life: Every living thing takes in energy and expels waste, maintains a thermodynamic balance, and, most remarkably, uses the same twenty amino acids to form the proteins that do all the heavy lifting within cells. What's more, all life on Earth uses the nucleotides adenosine triphosphate (ATP) and adenosine diphosphate (ADP) to store and distribute energy within cells. It's been extraordinarily difficult—somewhat like imagining a new color in the spectrum—to put together any respectable theories of what a significantly different extraterrestrial life might look like and how it might work. The National Academy report on "weird life" put it succinctly: "As Carl Sagan noted, it is not surprising that carbon-based organisms breathing oxygen and composed of 60 percent water would conclude that life must be based on carbon and water and metabolize free oxygen."

So any discovery of a life-form not descended from LUCA would provide this remarkable bonus: Finding Extraterrestrial Life 2.0 would make far more clear what Earthly Life 1.0 actually entails, and what is needed for something to be alive. It's a head-spinning conclusion, counterintuitive in the extreme. But the discovery of extraterrestrial life, of life as we don't know it, may be the only way to finally define what actually constitutes life on Earth.

4 THE SPARK OF LIFE

This is the story of an iconic but largely dismissed experiment that suddenly came back after fifty-five years to offer clues to a new generation of scientists about the spark that may have ignited life on Earth.

The most famous of all origin-of-life experiments was conducted in 1952 by University of Chicago graduate student Stanley Miller and his professor, Nobel Prize winner Harold Urey, in a small chemistry lab at their school. The experiment consisted of mixing together heated water with the gases believed at the time to have made up the atmosphere of the early Earth, and then exposing them to a prolonged electric shock. It was done inside a closed circuit of glass tubes and flasks hung on a metal scaffold, with no computers to drive and monitor the process, no gas chromatographs for high-tech analysis. But the results in its day were jaw-dropping: It showed that complex amino acids, essential organic building blocks of life, could be formed out of simple and common gases, water, and electricity. Prior to Miller-Urey, divine creation or the spontaneous generation of life were widely believed to be the only possible explanations for life on Earth. Darwin himself had dodged the issue in *On the Origin of Species.* Now science had demonstrated exactly how some of the key molecules needed for life could have come into being through natural processes, making plausible overnight the theory that life came together over eons in the early Earth's great primordial vat of oceanic soup.

The results were published at an especially heady time—just three weeks after James Watson and Francis Crick described the double helix structure of our DNA. That landmark discovery ignited the field of genetics and quickly led to new insights by the score, including the fact that our biological inheritance is passed on through genes that code the twenty amino acids essential for life. These same twenty molecules in turn provide the design for the workhorse proteins that actually make things happen in a cell—quickly breaking down compounds for energy and building new useful ones, creating a scaffolding within the cell, an immune system, and an ability to reproduce. And it's not just we humans that need these particular twenty amino acids (out of a possible universe of five hundred or more). These twenty compounds are also found in every living creature ever studied. Amino acids are not themselves living, but they are an essential organic component of proteins and therefore life, and the Miller-Urey experiment proved they could be formed, through tried-and-true chemistry, from inorganic parts. It was the beginning, in concept at least, of what would later be called "abiogenesis," "exobiology," and finally "astrobiology."

The experiment gave support to those looking for an entirely naturalistic and secular explanation for the origin of life on Earth, but it also did the same for those thinking about life beyond Earth. As Carl Sagan put it, the experiment was "the single most significant step in convincing many scientists that life is likely to be abundant in the cosmos." If life could be painstakingly assembled from nonliving things on Earth via the known laws of physics and chemistry, then why wouldn't the same processes produce life elsewhere?

But Miller-Urey didn't fare as well, over time, as Watson-Crick. While Miller-Urey gave impetus and energy to the study of prebiotic chemistry and inspired many researchers to push further in the field, scientists gradually came to believe that the gases used to make the amino acids in the experiment were actually not present on early Earth (some 4 billion years ago) in the forms and concentrations the researchers used. What's more, the carbon dioxide and nitrogen now believed to have dominated the atmosphere of early Earth form nitrites, which destroy amino acids as quickly as they come together. Miller-Ureyites found that the nitrite problem was largely solved if large amounts of iron and calcium minerals were also present, but there's no evidence that was the case. The findings were consequently challenged and in time pretty much ignored. The Watson and Crick work became the foundation of the burgeoning fields of molecular biology, genetics, and genetic engineering.

And so it was without much fanfare that in 2005, after Stanley Miller died, his onetime graduate student and later collaborator at Scripps, Jeffrey Bada, began closing down and packing up Miller's old lab. His interest was in preserving important correspondences to and from his mentor and colleague, and also to search for interesting amino acids Miller may have collected. Bada later recalled coming across an old cigar box on a shelf, noting that it had some old and seemingly used vials in it, and then storing it somewhere without much thought. Only several years later did Bada learn what he had overlooked, during a fortuitous meeting with another origins-of-life colleague, Antonio Lazcano of the University of Mexico. The two were preparing to give back-to-back talks at a conference in Texas, and they went up to their hotel room to compare slides and images on their computers. They wanted to make sure their talks wouldn't overlap. Bada saw something most unexpected on Lazcano's screen.

"Antonio just started showing me a bunch of other stuff that he had on his computer, and all of a sudden—boom—up came this picture of this little vial," said the low-key Bada, suddenly animated. "I said, 'What's that?'"

"Stanley told me it was extract from one of his original experiments that he'd saved," replied Lazcano.

"I felt like he hit me over the head with a rock and knocked me on the ground," Bada continued. "I was so stunned, I said, 'What do you mean?'"

"Well," Lazcano said, "Stanley just reached up in his office on the bookshelf and got down this cardboard box and pulled out another little box and said, 'Yes, here's one of the portions of my first experiments.'"

It all then came back in a flash. Bada remembered taking that bigger box of old vials out of the lab, granting it no particular importance and, he hoped, putting it in storage rather than in the trash. Alternately very excited about what might well have been in the box and heartsick that he may have tossed it out, Bada raced with Lazcano back to San Diego. They sped to the lab and, with profound relief, almost immediately found the box—still filled with vials featuring Miller's careful writing and small bits of brownish tar, including some from his most famous experiments. The two had struck gold, and they knew it. Their mentor would be directing their scientific paths once again.

"I guess what he did was take a sample out [of each flask], dried it down, and stuck it in those vials and just figured, well, there is probably not enough in there to analyze, but I'll save it anyway," Bada said. "Today, with the modern analytical methods you have available, that little amount of resin blows our instruments off the scale."

The Miller-Urey experiment was back, fifty-five years after being first conducted, and would soon be opening the door once again to new insights into how the building blocks of life may have been assembled. Unknown even to his colleagues, Miller had conducted some parallel experiments using the same gases but, he noted at the time, in the kind of steam-rich, lightning-charged environment found in a volcano. Bada recruited two other former students in Miller's Scripps lab—Danny Glavin and Jason Dworkin—who had gone off to work at NASA's Goddard Space Flight Center, where they had available precision instruments Stanley Miller could only dream of. The two analyzed the results from eleven vials of residue from those "lost" experiments using their cutting-edge lab—overflowing with compound separation and detection equipment—and discovered that the lost-and-found-again samples had produced an even more impressive array of amino acids than Miller had ever detected, twenty-two in all. That second, unheralded experiment also involved this highly significant modification: It was designed to simulate the chemistry and dynamics that occur at the mouth of a volcano rather than in the early Earth atmosphere. Suddenly the Miller-Urey experiment was pertinent again, because it showed that essential precursor amino acids could have been forged in the mouth of a volcano on early Earth as now, even if the surrounding atmosphere was considerably different. An article was quickly published in the authoritative journal *Science* outlining the recovered results and the case for volcanoes as possible stovetops for cooking important ingredients for life. The article came out in late 2008, around the time that Mount Redoubt, one of many volcanic sites in a line from coastal Alaska out into the Pacific, was erupting with fireworks, lava, and a vast ash cloud. Bada saw footage of the eruption and was struck by the amount of lightning present.

Critics of Stanley Miller had often pointed to his use of an electric sparker as a highly implausible re-creation of the dynamics of early Earth. But, as it turns out, pioneering research into volcanic lightning is making his sparker seem not so crazy after all, and now a new generation of scientists believe volcanoes may offer up clues to the process of preparing the Earth for life.

Volcanoes spit out molten rock, water, and gases from deep in the Earth and, in the superhot cauldron of its mouth, chemical reactions can occur that would lead to formation of those otherwise absent amino acids, as well as transform some molecules into forms more conducive to biology. For instance, the tight triple bonds of the most common form of nitrogen, an element essential for life, are broken in the heat of a volcano and can then combine with other elements and form useful compounds. For the amino acids, the newly formed precursor compounds in the volcanic furnace would then undergo additional changes in the waters assumed in this scenario to surround the volcanoes (through a process discovered 150 years ago by German chemist Adolph Strecker) and would gradually emerge as fully formed amino acids. They would then settle in ponds and tidal basins, where they would get concentrated through the work of the sun and become available to someday be incorporated into RNA or DNA.

So with this new incarnation of Miller-Urey research in mind and Mount Redoubt predicted to erupt again soon, I headed to Alaska to meet up with a specialist in lightning. Bada and his colleagues are squarely in the world of astrobiology, but my lightning expert, Ronald Thomas of New Mexico Tech in Socorro, is not. He has spent years chasing after thunderstorms to better understand how and why lightning strikes as it does. He was able to interest the National Science Foundation in a most unusual proposal a few years ago, one that was inspired by his study of the Mount St. Augustine volcano, which erupted in Alaska in 2006. Lightning, observers have long known, tends to accompany volcano eruptions—Pliny even wrote about the lightning display at Mount Vesuvius when it blew and destroyed Pompeii. But research into how or why lightning might accompany eruptions has, by all accounts, been limited. So Thomas (who teaches electrical engineering) entered the field, and his findings could have real significance for the Miller-Urey legacy and the origins-of-life field.

During the Mount St. Augustine eruption, Thomas teamed up with Alaska Volcano Observatory scientist Steven McNutt, who had initially noticed the heavy lightning associated with the event. When Mount Redoubt was preparing to go off, McNutt contacted Thomas, who rushed up to Alaska's Kenai Peninsula to set up the antennas that would monitor the eruptions. His technique was simple but novel: He was going to collect radio waves from the erupting volcano, which would contain the information he needed about the lightning being discharged in and above the volcano. Especially early during an eruption, ash and rock can hide the lightning, and those early fireworks seemed to be the most interesting. Their team set up a Lightning Mapping Array with four stations two months prior to what turned out to be a delayed eruption. Each station, explained one team member, has "basically a simple TV antenna set to pick up channel three—the frequency that lightning radiates."

When the volcano first blew in late March 2009, the instruments set up by Thomas and his colleagues found that lightning flashes accompanied every single one of the more than twenty major eruptions that occurred over thirteen days. What's more, they detected an unusual kind of lightning in the early moments of the eruption: a "constant lightning" that contained an extraordinarily high number of short bursts of 40 to 100 feet. They've never actually been seen, but Thomas imagines they would dance like enormous sparks from a monumental spark generator. The energy being released was monumental, a mass of very large electrical sparks that, if similar to other recent observations of constant lightning, glowed orange-red at the peripheries and a deeper red close to the mouth. Thousands of small flashes each set off radio impulses that flew across the inlet and were captured by the waiting receivers, which then drew a picture in graphs of what was happening. "The really intense phase of constant lightning went on for twenty to thirty minutes," Thomas told me. "We saw more lightning than we'd generally see in a major thunderstorm." His fellow researcher McNutt said those lightning strikes at the start of the Redoubt eruption lasted only 1 to 2 milliseconds and were "a different kind of lightning than had ever been seen before." Quite obviously, a great deal of energy was being released, exactly what was needed to "spark" the gases and water erupting from the volcano and change them in ways that would produce amino acids.

The radio monitors set up by Thomas's team were all on the far side of robin's-egg-blue Cook Inlet, about forty miles away from Mount Redoubt. They had wanted to place them closer, but the combination of Redoubt's location on protected federal land and the obvious dangers of being too near the volcano made that impossible. By correlating the millisecond differences in their flashes and triangulating the distances from the radio monitors across the inlet, Thomas was able to map and measure the lightning strikes inside the ash clouds in a new way—showing where the lightning originated and how it spread. He had begun that kind of monitoring at the Mount St. Augustine eruption in 2006 and had seen some of the same phenomenon, but it was during the Mount Redoubt eruption that the "constant lightning" phase was confirmed and better characterized. For a man who has witnessed a lot of lightning, Thomas was impressed. "I'm not sure I've ever been so excited by lightning before," was his conclusion.

As the team waited for possible further eruptions, McNutt proposed a trip to the mouth of the still-steaming volcano in a small, six-seater airplane. We approached ten-thousand-foot Mount Redoubt from the back, and were promptly introduced to the volcano mouth, hissing and spitting a short distance from the peak. From some angles, the volcano still appeared to be smoldering inside with a yellow-orange glow. Not surprisingly, we tend to think of volcanoes in terms of the dangers posed by shooting rock and flowing lava, noxious smells, ash shower, and acid rain—the effects we can see on the Earth's surface and its creatures. But flying over the still-scalding mouth and the lava slide forming just below, volcanoes come across as the endpoint of a tortured pathway from the innards of the Earth up to the surface. They deliver molten rock as hot as 2,000 degrees Fahrenheit and high-pressure gases from a place that seldom enters our minds: the rocky mantle below the Earth's crust, where the frictions and pressures of the always jostling continental plates, combined with the heat emanating from the planet's liquid outer core, can become so intense they, well, melt rock. Even a clouded peek into that world is unsettling; the power is just so unimaginable. Getting an up-close glimpse of the power and magnitude of a volcano—even without its supercharged lightning bursts—made the kind of early Earth chemistry proposed by Miller seem entirely possible.

As we circled the volcano, we could also see a thin, cloudlike haze and plumes above the peak of what looked like a gigantic steam grate at the volcano's mouth. The haze, Thomas and McNutt explain, is sulfur dioxide (which erupts with the magma and can lead to damaging acid rain) and the plume is largely H_2O. Magma contains huge quantities of water—more than in a thunderstorm, Thomas said—released during the melting of the rocks. If the airplane windows were opened, we would also have smelled the rotten-egg odor of hydrogen sulfide and would have been bathed in methane and nitrogen as well. The violence of a volcano's eruption is largely a function of the gases present; through their natural expansion, they determine just how explosive the eruption will be based on what gases are present and in what quantities they concentrate. But the gases that emerge are not necessarily the gases that settle on the surrounding landscape. The great temperature and pressure changes at eruption and all those lightning bolts of highly charged electricity set off chemical reactions that modify or transform the gases. That's the cascade of changes that Stanley Miller identified decades ago, and that Bada and his colleagues recently found to be even more enhanced in Miller's initially unpublished but now available experiment designed especially for volcanic conditions.

Earth has always had volcanoes of all kinds and in all climates: Volcanologists estimate there are about six hundred active ones today, and believe there were many more in the distant past (and on other planets like Venus and Mars). Some of those Earthly volcanic ranges, including the ones that formed the Hawaiian Islands and Iceland, bring up gases more directly from the scalding outer core, lessening their time in the cooling mantle. They also provide the kind of protective micro-environments that Bada and Lazcano believe played a major role in the origins of life, when much of the globe was covered in oceans only dotted with volcanic islands. The building blocks for the formation of amino acids and then proteins were all there, and Miller and his disciples showed in the lab a pathway for the process to work. We will never know for sure how that initial transformation from nonliving to living occurred because the evidence is gone. But we can know how the building blocks were formed from simpler, more elemental molecules.

Several months after Mount Redoubt calmed down, Bada came across some other potentially interesting vials in the old Miller cache. The one that most intrigued him was the brown residue produced by another experiment in 1958, one that used different gases than the original iconic Miller-Urey experiment. The 1958 model eliminated the hydrogen and added carbon dioxide, thereby responding to critics who said Miller was using the wrong gases to replicate atmospheric conditions on the early Earth. Miller also added the foul-smelling hydrogen sulfide that generally accompanies volcanic explosions and again sparked the mix with lightning-like electricity. The results, Bada said, were even more remarkable than the earlier experiments—with a greater number of amino acids produced and a better lineup of isomers, compounds with the same molecular formula but with different structures.

"I stand back and look at what's happened and I'm still dumbstruck," Bada said. "We've had the results of these experiments for at least ten years yet we never knew they existed. It's incredible new information and really shows Stanley's influence is with us today."

Antonio Lazcano, Bada's longtime colleague and current collaborator in testing more of Miller's vials, added this footnote: Asked why Miller never published the robust results from the experiment that added the hydrogen sulfide, Lazcano gave a very human, if scientifically questionable, reply. "Stanley always said he hated the rotten-egg smell of the hydrogen sulfide," he said. "I think he just didn't want to be around it."

The legacy of Stanley Miller is now very much living again not only because of the newly found volcano results, but because he trained men like Bada and Lazcano, who in turn have trained or mentored biochemists and astrochemists such as Glavin and Dworkin. They are the golden boys of NASA's Goddard Space Flight Center, located outside Washington, D.C., who used their cachet to lobby for a new generation of high-powered instruments that can analyze chemical and biological specimens—including the rediscovered Miller-Urey material—and map their electromagnetic fingerprint (and thus their chemical makeup) with greatly increased precision. They use their new tools to study the building blocks of life in meteorites, in comets, and in the vastness of space, and they have taken the origins-of-life field in some surprising new directions. One of their focuses has been on the peculiar dynamics of chirality—the uniform and unusual "handedness" of proteins and sugars—a phenomenon as puzzling as it is potentially revelatory. It's a tricky concept to take in, but basically involves the logic of mirror images: entities with the exact same structure, but set in an opposite direction.

Scientists have worked for decades to understand chirality, which has applications in drug-making and could someday be important in determining whether a microbe found on Mars or Europa or a comet is related to life on Earth or has an entirely separate origin. But it's virtually unknown to most everyone else. Here's a quick introduction:

In the mid-nineteenth century, the renowned microbiologist and chemist Louis Pasteur took on one of the puzzles of the day: why a solution with tartaric acid derived from living things (in this case, the discarded yeast remains of the winemaking process) behaved differently from a solution with the same tartaric acid that had been made synthetically. The difference involved how light passed through the liquid. The light entering the biologically derived solution didn't go straight through but rather was refracted off to one side, while the chemically synthesized sample allowed the light to pass through unchanged. Pasteur grew crystals of a compound including the tartaric acid and noticed that they came in two nonsymmetric forms that were nonetheless mirror images of each other. He laboriously separated the two forms of the compounds and found that solutions of one form rotated light clockwise while the other form rotated light counterclockwise. If the solution had an equal mix of the two versions of the molecule, then there was no effect on the light at all. What Pasteur correctly made of this is that the molecules of tartaric acid could exist in either a "right-handed" or "left-handed" form, rather like left- and right-handed gloves. The left-handed version was always biological, the right-handed one always synthesized.

I met Glavin and Dworkin one morning to learn about the unlikely role of chirality in astrobiology. They are astrobiologists now, but have backgrounds in physics, planetary sciences, and astrochemistry. Our destination was the Smithsonian's National Museum of Natural History and its collection of meteorites in a gallery called "Geology, Gems and Minerals." To be precise, meteorites don't fit any of those listed categories. But among the incoming rocks on display are the Allende meteorite from Mexico, the Orgueil meteorite from France, the Murchison meteorite from Australia—all legendary in the world of meteorites, and of especially great interest to astrobiologists. You couldn't tell by looking at it, or by gauging the pedestrian display or reading the brief description, but the dark gray Murchison meteorite in particular has become something of a holy grail for astrobiology. That's why Glavin, Dworkin, and other researchers come to the museum to collect precious samples stored away in a collection room closed to the public and filled with bagged and preserved pieces of Murchison and other rocks arrived from space. Of all the meteorite samples in the vast Smithsonian collection, about one quarter of the requests for samples are for a small piece of Murchison.

Pieces of Murchison hardly look remarkable now, and would be easy to ignore if you passed them by on the ground—just another faded dark gray rock with some shinier white highlights inside. But when it burned through the atmosphere in 1969 and landed in the Australian countryside, it was a gift from beyond. The roughly two hundred pounds of meteorite, broken up into rocks large and small, were collected with unusual speed, and in some cases protected from contamination. (In other cases, contamination is a greater issue—including the pieces that fell in a barnyard and were tossed by the farmer into the microbe-filled manure heap.) In the years that followed, hundreds of scientific papers have been written based on the ancient geochemical treasures found inside Murchison.

What makes it so important is that it is an extremely rare meteorite with lots of organic carbon (as opposed to the many others made up primarily of iron, nickel, silicon compounds, or inorganic carbons like graphite), and so is an extraterrestrial laboratory for the kind of carbon-based chemistry that ultimately produces life on Earth. To the initial amazement and delight of researchers, Murchison was found to be loaded with some of the important and complex molecules needed for life—including some of those amino acids that are used to make proteins. Just as English words are made up of the twenty-six-letter alphabet, all proteins—the workhorse molecules of life—are made up of twenty of the hundreds of amino acids known or suspected to exist. About half of those twenty amino acids essential for biology on Earth were found in Murchison. What does this mean? It does not mean what may first come to mind—that the presence of those biological amino acids says the rocks once were home to living things. Rather, it's the fact that *only* half of the twenty protein amino acids have been found. The absence of the rest has been used to bat down charges that the meteorite was substantially contaminated by Earthly microbes. If the rocks had been widely contaminated, then all the Earth's biological amino acids would be present. In addition, Murchison has scores of other amino acids found on Earth but unassociated with life and many others with no known Earthly counterparts. All of those complex molecules, it had to be assumed, were formed from simpler elements while on asteroids, comets, or in the primordial cloud of dust and gas where the solar system was born—making Murchison a true Rosetta stone sent down from space.

Here's where it gets really interesting to Murchison researchers. On Earth, virtually all living cells containing amino acids (and therefore proteins) are entirely "left-handed." Virtually all have sugars of a molecular structure deemed to be entirely "right-handed." This works great for biology, because the proteins and sugars need to be of opposite structures to successfully interact. But this kind of "homochirality," with virtually all proteins and sugars structured entirely in the same way, is unique. Everything else that isn't biological on Earth contains molecules that are mixed right-handed and left-handed in roughly the same proportion. Rocks, water, inorganic chemical solutions—they all have molecules that are mixed right- and left-handed on Earth and, as far as we know, everywhere else. Yet somehow the proteins and sugars essential to life are significantly abnormal in a way that makes biology possible. Dworkin put it like this: Biology would still work fine if amino acids were all right-handed rather than all left-handed, but wouldn't work at all if they were mixed left and right. "If you mix them, life turns into something resembling scrambled eggs—it's a mess. Since life doesn't work with a mixture of left-handed and right-handed amino acids, the mystery is how did life decide—what made life choose left-handed amino acids over right-handed ones?"

The mystery remains unsolved, but scientists have been finding clues to help explain it. And a primary repository of those clues is the Murchison meteorite. The pioneering work was done in the mid-1990s by two biochemists at Arizona State University, Sandra Pizzarello and John Cronin. The two isolated one particular amino acid called isovaline from Murchison, concentrated it, and then did an analysis of its chirality. Since everything nonbiological was understood to have molecules equally left- and right-handed, the assumption had to be that the isovaline would also be equally mixed. But it was not. The imbalance wasn't great, but there were about 8 percent more left-handed isovaline molecules than right-handed ones in Murchison. The discovery opened the door to several remarkable possibilities: that forces in the cosmos could transform the handedness of some molecules traveling through space on meteors, asteroids, or even dust, and that the handedness could then be gradually exaggerated on Earth until—in the case of the twenty amino acids essential for life—that small excess would become complete left-handedness.

Glavin and Dworkin used their more sophisticated instruments a decade later to make similar measurements of isovaline from Murchison and other meteorites. What they found was similar but more pronounced: a left-handed excess of 18 percent. Since no other nonbiological molecules on Earth are known to have this kind of excess, the two concluded that the left-handedness of amino acids, proteins, and therefore life on Earth most likely came from beyond. That hypothesis is now broadly accepted.

Meteorites bombarded the Earth 4 billion years ago and they delivered an impressive amount of carbon-based material to Earth, something on the order of hundreds of thousands of tons per year. The infall is much smaller now, on the order of 5,500 tons per year and most in the form of micrometeorites or even smaller particles of interplanetary dust. So how did those space-faring amino acids develop their slight excesses of left-handedness? Long-term exposure to the cosmic radiation of neutron stars is the most likely explanation, since chirality has been shown to change slightly when exposed to certain kinds of highly energized light. But no firm consensus has yet produced an answer.

All of a sudden, the undistinguished chunk of Murchison on display at the museum looked not so homely to me anymore. A gray-black, alternately smooth and craggy rock the size of a baked potato and mounted on a simple blue rod, it has an improbable and important story to tell. Murchison, it is commonly believed, is somewhere between 4 billion and 4.5 billion years old—meaning that it was formed around the time that Earth was. It was initially a part of the asteroid belt between Mars and Jupiter, and no doubt was spun in the direction of Earth after another asteroid crashed into it. At some point it was exposed to water. Ironically, it can tell us more about the early history of the universe than anything on Earth because the planet has been so altered by oceans, volcanoes, the cycling of elements like carbon, and, most important, by life.

Glavin and Dworkin walked me down to the museum's meteorite collection, a treasure trove of gifts from outer space. Curator Linda Welzenbach got us gloved and began to take out more and more samples of Murchison, some in glass vials, some in plastic bags. In all, the museum has about sixty-five pounds of the meteorite, purchased by NASA from private collectors in 1971, two years after it fell. They sit in covered shelves inside a heavy-duty storage cabinet, hopefully not gathering dust. Welzenbach didn't hesitate when asked if we could take a look at some of her collection.

Holding pieces of Murchison was an unexpectedly moving experience. To be in physical contact with something so profoundly and undeniably old was to be, for one transporting moment, connected to that near eternity, to be drawn back to the Big Bang, the great mystery that science can now describe but not really explain. No wonder meteoriticists, as they're called, are a famously intense crew. They're in daily touch with the concrete reality of deep space and deep time, and their job is to understand what the rocks are telling them about both.

"There is certainly a constant debate in the origins-of-life community about what fraction of the organic material on the early Earth was made in situ, locally, by reactions on Earth versus extraterrestrial input," said Glavin. "A lot of people argue the amount coming in from [extraterrestrial] forces—meteorites, interplanetary dust—wouldn't be enough . . . to get good chemistry to take place. You could argue it either way."

"But the problem is the evidence is lost," Dworkin interrupted. "You can infer things based on modern observations of meteorites and based on laboratory experiments of what the chemistry might have been like. But the record is gone. And the best you can do is make intelligent guesses and see what's out there." Glavin finds it paradoxical that no clear record exists of ancient, prebiotic chemistry on Earth, but that clues into the nature of those precursors to life arrive regularly from outer space. And so while trained and still immersed in the Miller-Urey tradition, Glavin and Dworkin nonetheless focus on meteorites and comets rather than the origins of the Earthly ingredients that might once have made up the primordial soup. A recent discovery of theirs, for instance, involved an Earth-bound meteorite that was tracked in space, landed in northern Sudan in 2008, and was collected in more than 600 pieces within weeks. The rock was initially formed when two asteroids collided and caused a heat shock of up to 2,000 degrees Fahrenheit, which is considered to be well beyond the point where all complex organic molecules (including amino acids) should have been destroyed. But Glavin and his team found the amino acids there anyway. Having determined that Earthly contamination was not involved, Glavin concluded the meteorite is telling us the very important news that there appear to be numerous ways to form the seemingly abundant amino acids in space, a dynamic that "increases the chance for finding life elsewhere in the universe." Just as those amino acids fell on Earth and may well have played a role in the origin of life here, so too would they be falling on Mars, Europa, and planets and moons throughout the cosmos.

The initial Miller-Urey experiments set off an explosion of related research around the world, and scientists now know significantly more about the mechanics and logic of how nonlife might become life. But more than a half century after Miller-Urey, science remains without an explanation of the actual pathway, and many scientists no longer believe we will ever know the specifics of how life began on Earth. The original experiment was a spectacular proof of concept that was eagerly embraced by researchers, but the actual chemistry remains largely unraveled. Nonetheless, Miller-Ureyites have dominated, enhanced, annoyed, and frustrated the field ever since.

As described by British-born Mark Russell, a senior research fellow at the NASA Jet Propulsion Laboratory and a specialist in geochemistry, they have "commanded the high ground for fifty years" and, he argues, have hurt origin-of-life science. A brief 1988 critique of the prevailing Miller approach written by Russell and published in *Nature* resulted in a torrent of requests for reprints, and convinced him and colleagues that others bridled under the weight of the legacy. Russell's own work has been focused on undersea origin-of-life chemistry, the field created when the powerful hydrothermal vents called black smokers were first found and explored on the floor of the Pacific in the late 1970s and '80s. Surprising life-forms exist around the nutrient-rich heat of the smokers, leading to some theories that life started in their vicinity. Russell has long worked on a corollary to that notion, arguing that the smokers themselves were too intense to nurse life into being, but that the many subsidiary and less torrid undersea vents were ideal nurseries and provided the right kind of chemistry, geology, and potential energy. He was conducting a laboratory test of his theory when I met him at the Jet Propulsion Lab, located at the California Institute of Technology in Pasadena, California. But his position was temporary, and he wondered if the influence of the Miller-Urey progeny was part of the reason why.

Not only did the science of the Miller lab command the "high ground," Russell said, but it and its graduates also "commanded a lot of the NASA think tanks and commanded where their ex-students went to in terms of different universities, and generally kept control of origin-of-life science and evolution of biosphere." He said that as editors and peer reviewers for the major science journals, they also had inordinate control over what got published in those most prestigious publications. But fortunately there are other journals eager for his work, so the debate continues.

But even if the Miller-Urey details and chemical pathways remain debated, the basic validity of "abiogenesis" does remain in place—the assertion that the building blocks of life, and so life itself, can be formed from entirely nonbiological sources. This logic is still being used and expanded, and in some unusual places. There is probably no better example of this than the experiment commonly called "Miller-Urey in space."

The presence of organic compounds in space was first detected eight decades ago, but an understanding of their ubiquity is relatively new. It turns out they are most everywhere in space—carbon dioxide, more complex amino acids, and even formaldehyde. The most frequently found are polycyclic aromatic hydrocarbons (PAHs), an early phase of chemical evolution of carbon and hydrogen into the more complex molecules that are useful to life. Just as Miller and Urey wanted to learn if complex amino acids could be formed in the lab by sparking certain gases and water, a team of largely European researchers proposed a while back doing the same thing in space. The experiment was quickly approved for the orbiting International Space Station or ISS, though not so quickly designed, built, or scheduled for delivery. But it is now far along in development and remains a priority for the European Space Agency, and so "Miller-Urey in space" will very likely one day join Miller-Urey in Chicago as an important pathfinder in the origins of life.

But what a difference fifty years make. While Miller-Urey was famous for both its results and its remarkable simplicity, the effort to test similar dynamics in space is anything but simple. It involves shipping the experiment up to the space station and installing it into the Microgravity Science Glovebox in a pressureless, weightless room on board the European-built Columbus module of the station. The Glovebox will simulate low-temperature conditions at the solar nebula, the disk that formed after the collapse of an enormous cloud of dust and gas formed our solar system. Astronauts will place two sets of vials in the chamber with ice-crusted grains of silicon minerals and two different mixes of gases: one corresponding to the initial Miller-Urey experiment and the other taking into account current thinking about what was present in the solar nebula. Each vial is equipped with two electrodes that produce an electric charge, and the ice-covered particles float through the gaseous and electric mix, just like in the solar nebula. After being exposed for several weeks, the vials have to be returned to Earth in –36 degree Fahrenheit containment for analysis using instruments Stanley Miller could hardly dream of back in 1953.

The person heading the effort is Pascale Ehrenfreund, one of the stars of international astrobiology. She has been principal investigator or co-investigator on many NASA and ESA experiments, including satellites, planetary probes, and experiments on the International Space Station. The basic logic of the Miller-Urey experiment remains the same: to see whether combinations of gases, including water, can produce the nonliving but essential building blocks of life when hit with an electric charge. The difference is that Miller-Urey in space will do the experiment in zero gravity, and will be looking to mimic the formation of complex organic molecules in the solar nebula rather than the oceans of Earth.

Ehrenfreund, an astrochemist, is an expert in polycyclic aromatic hydrocarbons, which are known on Earth as a pollutant formed by burning fossil fuels but in astrobiology are important, much-studied players. In recent years PAHs have been detected across the universe, and are now known to be ubiquitous and plentiful, perhaps the most common complex chemical compound in interstellar space. They also make up about 70 percent of the organic carbon that Glavin and Dworkin study in carbon-based meteorites, micrometeorites, and interplanetary, or interstellar, dust that falls to Earth.

"So for us, an important goal of the experiment is also to see about the formation of polyaromatic hydrocarbons," Ehrenfreund told me. "They were falling onto Earth in large amounts when life was being formed, so obviously we want to know if they were somehow involved." At the very beginnings of the process that later resulted in life, she said, "everything had to be extremely simple and extremely robust to withstand the hostile conditions." PAHs are both, and also have a structure not too different from that of a cell wall, leading some to speculate that PAHs played a role in organizing cell structure, too.

Miller-Urey in space is an important experiment for Ehrenfreund, but it isn't her only work now flying on the ISS. She is a co-investigator as well on an experiment that exposes spores, bacteria, fungi, and other organic material to the harsh space environment by hanging them out of the Russian and European laboratories attached to the station—research designed in large part to test the theory of "panspermia," that life on Earth arrived from Mars via space. Results so far show that some microbes can indeed withstand the radiation, desiccation, and cold of space.

Ehrenfreund also leads an ISS experiment that dangles a metal container filled with PAHs and other complex carbon molecules from the space lab. The goal of this "Expose" experiment is to understand the effects of zero gravity and ultraviolet and other cosmic radiation on PAHs, and it focuses especially on photochemistry, the changes that occur when different forms of light hit the compounds. The Miller-Urey breakthrough involved learning how amino acids could be formed in the early days of Earth; scientists hope that this Expose experiment, as well as Miller-Urey in space, will help show how the compounds ultimately found in life came to be formed in the early solar system, billions of years before and many light-years away.

If you step back for a moment, you can see that this work is all part of an interplanetary and interstellar forensic quest to tease apart how the building blocks of life came to be on Earth. Since those same building blocks are landing on planets and moons around the universe, the obvious question is whether they are being used as the chemical backbones to life on other planets and moons, too. It's demanding, technically complex, and totally absorbing—solving mysteries that make crime forensics look, well, elementary. Since the Miller-Urey experiment and its primordial soup became household words a half century ago, nothing in the world of origin-of-life research has made anything like that kind of splash. Half joking, half serious, Ehrenfreund imagined how that might change.

"What we need is a movie or a TV show, something about how we're trying to solve these big mysteries," she said with a pleased laugh. "That's it—a show all about the forensics of life in the universe. I think it could be a hit."

5 ON THE TRAIL OF LIFE ON MARS

The scientist who first confirmed the presence of the gas methane on Mars at specific locations and at specific times—the best sign yet that the planet is, or was, alive—has been a rebel since he was a boy. Michael Mumma grew up in Amish country outside Philadelphia, Pennsylvania, in a family whose life was built around a farm and the fundamentalist Evangelical United Brethren Church. But Mumma had a strong independent streak—he liked to spend his days roaming the woods and fields and at ten he took on his own newspaper route to build an income stream of his own. At eleven, he confronted his pastor. The science teacher had told his class that the Earth was more than 4 billion years old, that humans were the product of evolution, and that a fossil record existed that supported both ideas. So the world was not created by God in seven days? "My minister had a simple response when I asked him about it," Mumma recalls. "He said that to be part of the church, I had to believe what is taught here and not what the teacher says. He said the teacher at school was wrong."

More than five decades later, you can pretty much draw a straight line from that encounter between Mumma and his minister and the announcement in early 2009 of one of the most significant discoveries in astrobiology: the appearance of methane gas on Mars at predictable spots and at regular intervals. Now Mumma is on a quest to understand how and why the gas is spewing out from the Martian underground, and what its origin might be. On Earth, some 90 percent of the methane in the atmosphere is a by-product of living creatures, and biology has to be considered a serious candidate for the production of methane on Mars as well. In other words, the methane could be the result of living or once-living extraterrestrial life. It is also true that methane can come out of a geological process. But the alternative to biology would be almost as significant: Mars has long been considered geologically dead—without volcanoes, earthquakes, and submerged moving plates—so the presence of geologically replenished methane would mean it is not.

That straight line running from Mumma's early disenchantment with his minister to his Mars announcement is drawn like this: Mumma's already strong independent streak was given a powerful boost by his encounter with the minister and the break with his church that followed. He later became the first in his family to leave Lancaster County and the fold since his ancestors arrived there in 1731. But rather than defeating him, his departure put him on the path to becoming a seasoned researcher known for his mantra of always following the scientific data and his hardwired habit of thinking outside the box. That willingness to buck the community—be it religious or scientific—was essential, because he could never have made the methane discovery had he stuck to tried-and-true strategies, and had he been swayed by the many people who told him his approach was misguided and his goal most likely unattainable.

Mumma first came to Goddard, in Greenbelt, Maryland, just outside Washington, D.C., in 1970, choosing it over six other job offers at a time when students with doctorates in atomic and molecular physics were not exactly in high demand. He set to work on improving—revolutionizing, really—the use of spectroscopy in astronomy, the gathering of photons from afar and breaking them down into distinguishable component parts. His particular job was to build a spectrometer that could better detect and identify the molecules in comets, using the signatures of particles coming off the nuclei of atoms. He and colleagues at NASA Goddard came up with the technology to take readings of distant molecules in the previously little-used infrared band, a high-energy form of radiation on the electromagnetic spectrum. Using the infrared allowed Mumma to collect more subtle data that could much better identify and define distant molecules and compounds, a process akin to building a telescope with a bigger mirror that can see farther and with more precision. Without the ever-increasing detection power of infrared spectroscopy, Mumma could never have identified those seasonal releases of methane on Mars, nor would he have had the tools needed to determine whether the methane comes, at least in part, from some living, or once-living, creatures.

The great excitement within astrobiology about methane on Mars comes not only from the discovery of the gas, but also from where on the planet and when in the Martian year it was found. So far, the methane (a carbon atom surrounded by four hydrogen atoms) has been detected at five locations, and all have at least one thing in common: They are areas where the remnants of early Mars, that is, when it was most Earthlike and most hospitable to life, are least disturbed. Specifically, that means a time when liquid water appears to have run freely around the planet, and when Mars had a magnetic field surrounding it that enabled a much thicker atmosphere to act as a shield against the ravages of the solar wind and the ultraviolet radiation that now desiccate the surface. This was some 3.5–4.5 billion years ago in what is called the Noachian age of Mars, yet remainders of those days have been detected in low levels of magnetism still present in some areas. Mars scientists have also found evidence of minerals that can only be formed and transformed in the presence of liquid water. In other words, the methane seems to be coming from areas most protected from the yet-unknown catastrophic events that turned Mars from a watery planet shielded from the harsh forces of the sun and cosmic radiation to the parched landscape of today, with its thin atmosphere unable to ward off those formidable assaults.

On Earth, much of the methane now produced comes from cows, forest fires, and landfills. But the earliest biological source was no doubt methanogens—primitive microorganisms that both produce and consume methane, and have been here making their "marsh gas" since earliest times.

They appeared before oxygen was abundant in Earth's atmosphere, and so the dearth of oxygen on Mars would be no problem for these tiny round or rod-shaped organisms; they can survive for some time in the presence of oxygen, but generally live without it. In theory, then, similar methanogens could be doing the same on Mars near the surface or even far below, and they could be doing it on other planets or moons as well. The Mars methane release would, in this scenario, occur when the right climatic conditions arrive for an explosion of the methanogen population, or it could come from a melting of surface ice that then allows subterranean, bacteria-produced methane to escape. In other words, during the relatively warmer Martian summer.

But there are also two geological contenders as the source of the methane. The first involves a process called serpentinization, where water and carbon dioxide interact with particular minerals in a way that makes methane as a by-product. The mineral most subject to this serpentinization process is called olivine—also known in its gemstone form as peridot—and it is common on Mars. The second possible geological source is erupting volcanoes. Methane is known to accompany lava as a volcano erupts, but on Earth the gas sulfur dioxide is always present as well. So far, sulfur dioxide has never been detected in the Martian atmosphere. A final possibility is that huge reservoirs of the methane remain deep underground and leak slowly up. They may have been produced biologically, geologically, or as part of the process that brought Mars into existence some 4.5 billion years ago.

But what is known for sure is that the methane releases occur regularly, because the gas doesn't stay in the Martian atmosphere for long. On Earth, methane produced by bogs or cows can remain for a decade; on Mars, for reasons that are not fully understood but imply a photochemical destruction process unlike anything now known, the gas is destroyed in under a year. Identifying that short lifetime was one of the major accomplishments of Mumma's team because it allowed them to break with the conventional wisdom—that there was little or no methane on Mars—and show that at certain places and at certain times, there are major releases. The amount of gas being released is substantial: about 20,000 metric tons of methane in the big plumes during summer in the northern hemisphere. That rate is comparable to the natural hydrocarbon seep at Coal Oil Point off Santa Barbara, California, which is the largest of its kind in the Western Hemisphere and the second largest in the world. The natural gas seepages may well be the result of different dynamics on Mars and Earth, but Mumma considers the comparable sizes to be significant.

Determining how these plumes of Martian methane were being produced would seem impossible from Earth, yet there are promising theories about how to do this. They involve observations and calculations that are stunningly complex, and their absolute accuracy probably won't be known until astronauts land on Mars and perform tests that can only be done on the surface itself. But Mumma has put together what he describes as a campaign to accelerate his breakthrough work. NASA and many of the major observatories in the world are now eager to support his team with an almost unprecedented twenty-three observing runs on the world's most powerful telescopes in the next few years.

Potentially even more important, the methane discovery abruptly changed both American and European plans for the next generation of missions to Mars. The previously planned European ExoMars expeditions of 2016 and 2018 became a joint NASA-ESA mission within months of Mumma's methane announcement, and searching for methane became a major—probably *the* major—focus. The missions will feature an orbiter equipped to locate the most interesting gas releases, and then the deployment of landers two years later to explore the best of those sites. The explicit goal will be to search for biology on Mars, something that hasn't really happened since Viking in 1976. Mumma's methane discovery probably saved the NASA Mars program from a significant downsizing, and the European Mars mission as well.

Mumma's recent campaign began in late summer of 2009 in the Atacama Desert of northern Chile. The Very Large Telescope observatory at Paranal, built and operated by the European Organisation for Astronomical Research in the Southern Hemisphere (ESO), sits atop a desert mountain, with four big telescopes at the peak and eight smaller ones nearby. The area is the copper belt of Chile, where much of the nation's wealth is pulled from the ground. But not around Paranal—for at least thirty miles in all directions there are no people at all, no animals, no trees, hardly any plants or even insects. Once out of the environs of the sprawling, gritty port town of Antofagasta, what you see are bare hills and coarse, sandy plains. To encounter a boulder—or better yet, a field of boulders deposited by fast-running water last seen hundreds of thousands of years ago—is about as visually exciting as it gets. By many measures, this is the driest place on Earth, or at least among the top three. The harshness of the area was what drew ESO to Paranal in the mid-1990s, because environments that support plants and animals generally bring people, and people produce nighttime light. Operating a complex, high-precision set of very expensive telescopes is hard enough on its own; doing it surrounded by towns, villages, and their light makes the job much harder. Terrestrial light is definitely the enemy at Paranal, and fortunately there are eighty miles between the telescopes and the bright lights of Antofagasta.

The road to the observatory breaks off from a newly paved highway after some ninety minutes of desert driving. The observatory road loops over small hills and down to a gentle but equally parched valley with dozens of all-white buildings large and small. At first encounter they seemed evacuated. Only later did I learn that outside activity is discouraged during the day because the level of harmful ultraviolet light in the area is dangerously high. Everyone, as a result, was inside. My ESO guide, former astronomer Laura Ventura, parked beside a white prefab building; off to the left was a large, low-slung half-dome in the gravel. We stepped out into the piercing sunlight, Ventura told me to grab my suitcase, and we headed toward what appeared to be a futuristic maintenance building or storehouse. We came to a long, declining ramp that led us below ground level to a pair of heavy metal doors. We tugged them open and passed through to a small, pitch-dark room leading to another set of doors. I pulled open the second door and it was as if we'd been swept suddenly and miraculously away to the Amazon or Costa Rica or any evergreen equatorial rain forest. Outside was the driest desert on Earth, but inside was the tropics, and it was bathing us in sultry, humid air.

Before us was a small, sloping jungle of full-grown palms, ferns, and banana trees, all under that half-dome, and all underground. A swimming pool shimmered at the bottom of the faux hillside. We had arrived at La Residencia, the hidden home of all who explore the heavens at Paranal. My arrival was accompanied by an admonition to keep my room windows covered by the blackout shades once the sun goes down, and not to worry if the ground began to shake—earthquakes in the area are common. I was introduced to the all-important wake-up call system that rouses researchers at all hours of the night and told to be up at two-thirty in the morning.

What brought Mumma, his Goddard collaborator Geronimo Villanueva, and former Mumma postdoc and now ESO "Infrared Instrument Scientist" Hans-Ulrich Käufl to Paranal was an unusually precise piece of equipment designed and modified with their methane-on-Mars work in mind. It was recently attached to UT1, one of the four eight-meter mirrors of the Very Large Telescope array atop the mountain nearby. At the time, the instrument—CRIRES, or the cryogenic high-resolution infrared echelle spectrograph—may have been the most powerful land-based astronomical instrument in the world for identifying and characterizing gases such as methane on distant planets and stars, allowing for a more precise locating of the methane plumes than ever before. Ironically, the team's most pressing concern before arriving at Paranal was that their target—Mars—might be too close for the instrument to produce good measurements. A telescope and spectrometer designed to see into distant space could have a hard time with an object as near as Mars.

The process of collecting photons from around the universe through the telescope, funneling them into a vacuum-sealed, supercooled (to –328 Fahrenheit) spectrometer like CRIRES, and coming up with potentially useful data is a marvel of precision. It requires a multitude of transformations, adaptations, reconfigurations, and judgments. That afternoon I was ferried to the mountaintop and introduced to the telescopes and the spectrometer, which is affixed to the base of the big mirror like an oversize parasite to its host. At its most basic, CRIRES operates under the same logic as a simple visible light prism: It takes in photons and breaks them down into component parts, to their equivalent of a rainbow. CRIRES works in the infrared band of the spectrum, but other spectrometers read radiation sources ranging from low-frequency radio and microwaves to visible "white" light and to higher-frequency ultraviolet and X-ray. What the instruments read is often the movement of electrons inside an atom or molecule that have been excited by starlight and other energy sources. The excitation makes the particles move in a detectable way that provides a recognizable signature in photons for one, and only one, element or molecule.

It's all a function of quantum physics, so it is both vaguely mysterious and remarkably precise. The result is that when Mumma and his colleagues search for methane on Mars, they are not using the kind of detection instruments you'd find in the oil and gas industry, or to identify a gas leak, or to monitor releases from landfills. And he's not looking through the lens of a telescope. Instead, he's reading spectra on a computer screen.

Scientists have known about the dynamics of the electromagnetic spectrum for two centuries, but the usefulness of spectroscopy for astronomy only became apparent in the early twentieth century. It began in 1859 with construction of the first spectrometer by German chemist Robert Bunsen (of Bunsen burner fame). In 1864, two amateur British astronomers—Sir William Huggins and his wife, Margaret Lindsay Murray Huggins—used a rudimentary spectrometer to discover that the sun is made up largely of hydrogen. Spectroscopy led directly to the 1868 discovery of helium by an amateur British astronomer, Norman Lockyer, who came across unusual spectral lines coming from the outer atmosphere of the sun during an eclipse. But it was only in the early twentieth century, after quantum physics and quantum mechanics revolutionized our understanding of how particles move and spin on the atomic level, that its full potential was revealed.

Elements needed for life on Earth—including carbon, oxygen, nitrogen, hydrogen—have been found throughout the universe. These star-forming nebulas (top: in the Sagittarius arm of the Milky Way; bottom: the Carina nebula) can produce complex carbon compounds that are the building blocks for life as we know it. If those compounds fell on exoplanets orbiting in the "habitable" zone around their stars, would they become part of the life-creating process that occurred on Earth? TOP: *NASA, JPL/Caltech/M. Povich;* BOTTOM: *Hubble Space Telescope, NASA, European Space Agency and J. Hester*

Deep in the Northam Platinum mine in South Africa, Gaetan Borgonie of Belgium's Ghent University hunts for extremophiles, the microscopic organisms that thrive in extraordinarily hot, cold, acidic, or otherwise inhospitable environments—like those on planets other than Earth. The temperatures inside the rock walls can exceed 150 degrees Fahrenheit. *Marc Kaufman*

Breaking into Antarctica's Taylor Glacier with chainsaws and demolition hammers, members of Brent Christner's team from Louisiana State University mine for extremophiles embedded in ice. The team dug 40 feet into the glacier, which towered 100 feet above them, to retrieve pristine samples of the ice. *Shawn Doyle*

A block of ice from near the bottom of the Taylor Glacier rests in a cold room at LSU. Christner brought back 3,300 pounds of the glacier's ice from his 2009 expedition and says living microbes are present in many of the samples. *Marc Kaufman*

The *Endurance* "Bot" emerges through a melt hole in Antarctica's Lake Bonney to the delight of Peter Doran of the University of Illinois, Chicago, and Vickie Segal of Stone Aerospace. Scientists hope that submersibles like this one, designed to explore and collect data autonomously under the ice, will one day explore the ocean under the icy crust of Jupiter's moon Europa. *Stone Aerospace, Austin, TX*

NASA Astrobiology Research Fellow Felisa Wolfe-Simon and her colleagues reported in late 2010 finding microbes in California's Mono Lake that could grow in an arsenic-rich medium and that appeared to use the usually toxic element in their DNA and cell membranes. The discovery may represent the first finding of life that does not rely on phosphorus as one of its six essential elements (along with carbon, oxygen, hydrogen, nitrogen and sulfur). *Henry Bortman*

Scientists have long debated whether the black-purple varnish that streaks the walls of the Gorge at Utah's Capitol Reef National Park indicates the presence of life or merely the result of chemistry. The dispute about "desert varnish" is a window into the larger issue of how to define "life"—if such a definition is even possible. *Janet Little*

Researcher Jeffrey Bada at the Scripps Institution of Oceanography reconstructed the iconic Miller-Urey experiment, after discovering long lost original samples from the 1950s. The experiment revealed intriguing new information about how the building blocks of life might have been formed on early Earth. Bada, a former student of Stanley Miller, sees volcanoes as likely furnaces for the creation of that primordial soup. *Marc Kaufman*

Alaska's Mount Redoubt erupted in 2009, and lightning experts from New Mexico Tech and the Alaska Volcano Observatory detected what they called intense but short bursts of "constant lightning" at the mouth of the volcano. That kind of natural electricity parallels the electric charge that Stanley Miller used in making amino acids—a building block of life—out of gases and water. *Bretwood Higman*

The Murchison meteorite fell in Australia in 1969 and has become the most studied meteorite on Earth. Murchison has been classified as a "carbonaceous chondrite," and researchers have found that it contains relatively high levels of amino acids, carbon compounds, and nucleobases—molecules essential to the workings of DNA and RNA. The presence of this chemistry in the meteorite raises questions about whether the building blocks of life—or perhaps life itself—came to Earth from afar. *Chip Clark, Smithsonian Institution*

The European Southern Observatory at Paranal, in Chile's Atacama Desert, is one of several arrays of very large telescopes that have made it possible to hunt for exoplanets and study distant atmospheres from Earth. *European Southern Observatory*

Kueyen is one of the four 8-meter-wide telescopes at the Paranal observatory. The telescopes and new spectrometers are so powerful that NASA scientist Michael Mumma was worried they couldn't observe an object as close as Mars. Both European and American teams are now racing to build 30-meter-wide or larger mirrors for the next generation of telescopes. *European Southern Observatory*

Using a telescope-based spectrometer at Hawaii's Keck Observatory, Mumma detected methane releases at particular Martian locales and at particular times of the year. A concentration map of the methane measured shows the gas coming from relatively undisturbed regions, which links the methane to Mars's early history. *NASA*

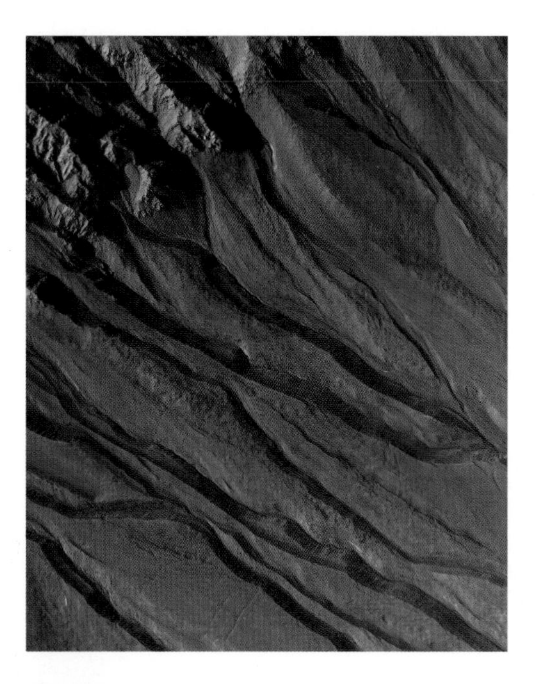

Planetary scientists are now convinced that Mars was once warm and wet, with running water that could have formed what appear to be ancient gullies. Mars was at its most habitable 4 billion years ago, and some theorize that life from Mars came via meteorite to Earth, which was then far less stable and conducive to life. *NASA*

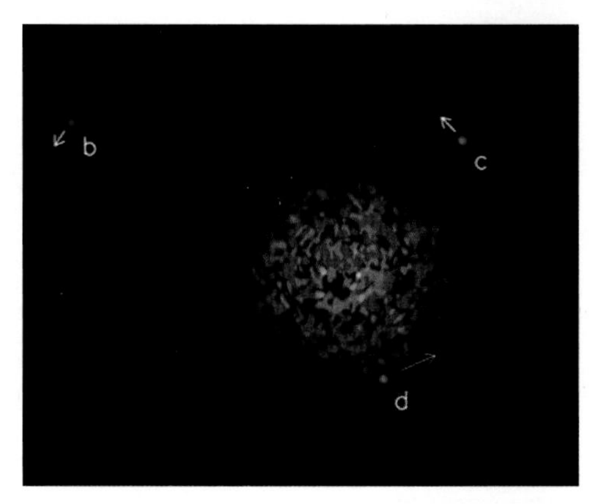

This first direct image of a star with three orbiting exoplanets was released in 2008. Most exoplanets are detected through indirect measures of the planets' effects on their suns, but ground-based infrared telescopes captured this image of star HR8799 in the constellation Pegasus, 129 light-years away. The planets are gas giants like Jupiter, only many times bigger. *NRC-HIA, C. Marois & Keck Observatory*

An array of forty-two radio telescopes at Hat Creek in northern California was built in part to search for signals sent out by intelligent extraterrestrials. The array, funded largely by Microsoft co-founder Paul Allen, allows SETI scientists to listen in a far more comprehensive way. But they caution that even now, of the 100 to 200 billion stars in the Milky Way, only a tiny fraction has ever been targeted. *Marc Kaufman*

What appear to be twinkling stars in this ultra deep field image taken by the Hubble Space Telescope are instead full galaxies thousands of light-years across. As scientists have come to understand the vastness of our expanding universe, they have also wrestled to comprehend the fine-tuning of the cosmos—the fact that many of its physical laws had to be almost exactly what they are to allow for the existence of life. Even a slightly different ordering would have left the universe without the carbon needed for life as we know it, without the ability for most chemistry to occur, and without stars. *NASA*

More recently, a spectrometer of the orbiting Hubble Space Telescope in 2001 detected the first element isolated from an exoplanet when it found sodium in the atmosphere of a planet fifty light-years away. Spectroscopy has since been used to measure the evaporation of hydrogen from the atmosphere of a planet 160 light-years away, and more recently the presence of carbon dioxide and methane surrounding a large gaseous planet 63 light-years away. Without spectroscopes, astronomy would lose half its ability to see into the cosmos.

Mumma has had a lot of experience with spectroscopy: His early work at Goddard focused on comets, then on the possibility of finding the organic gas ammonia on Mars. After that petered out, his search for methane began. The year was 1989, two decades before he finally went fully public with his findings. In the worlds of planetary science and astrobiology, discoveries generally don't come fast.

It is a tradition at Paranal to introduce newcomers like myself to the telescopes at sunset. Not only are the four enormous domes grand and sparkling in the low golden sunshine, but the mountaintop is said to be an auspicious place to see the "green flash." The flash is a very brief glimmer of green light that, for the lucky few, can be seen seconds after the sun has gone down over the horizon. It's a real optical phenomenon, as opposed to an optical illusion, that involves the refraction of light and the differing frequencies of blue-green light versus red-orange. It's best seen over the ocean, and we had a glorious view of the Pacific. Or so I thought. I had sensed that something was slightly off, that the water seemed closer than it should be, but nonetheless I was seeing waves head toward shore and I even saw "islands" in the water. It took a while, but it gradually became apparent that those waves were in the clouds and those islands were hilltops and mountain peaks. The huge, shoreline bank of clouds that typically rolls along the Atacama coastline fooled me, as it has many others. Perhaps it was because we were watching the sun set over Pacific clouds rather than the Pacific Ocean that the green flash didn't appear, but the sky did turn a bright orange and then an otherworldly bloodred. As the sunset played out to the west, the big domes behind us began to open, the telescopes were set into motion to a series of groaning, ghostly sounds, and the night's observations began.

But Mars wouldn't rise in the sky until early morning, so we returned to La Residencia for a brief sleep and that middle-of-the-night wake-up call to return to the observatory. But first some stargazing. From our high Atacama perch, the visible universe held an infinite collection of starshine. The Milky Way was not a distant smudge but a clear and vast disk of stars; the wispy Magellanic Clouds (two very near and star-packed galaxies) looked like so much celestial cotton candy; the supergiant star Betelgeuse (which gives off one hundred thousand times more light than our sun) glowed red. Reluctantly, we piled into several cars for the ten-minute ride up to the telescopes, built on a platform atop a flattened 8,650-foot peak. The drivers, of course, did not turn on their lights, so the sharp mountain turns demanded practiced care. We arrived alongside the four big scopes—named in the local Mapuche language Antu (the sun), Kueyen (the moon), Melipal (the Southern Cross), and Yepun (Venus)—and headed for a small side entrance of the first, then up two flights of stairs, around a bend, and into the control room. It was alive with computers and screens of all kinds, but not with many people. The telescope itself was also nowhere to be seen; its huge mirrors, located behind a thick wall in a cavernous chamber, take in photons and send them to various instruments to cut and slice as the researchers need. The Mumma team prepared its instruments and computers for the coming Mars-rise.

The first days of the campaign had been devoted to assessing whether concentrations of methane tended to be released alongside concentrations of water. The data on that had previously been inconclusive, but now the team would be able to simultaneously detect and map, via CRIRES, the methane and water at a particular location better than before, and for the first time to detect methane and its biologically important relative ethane. Simultaneously, they would examine the isotopic characteristics of the water (seeing if some of the hydrogen had an extra neutron) and would search for other hydrocarbons related to methane. But most important, the campaign would involve replicating and expanding on data about the presence and dynamics of those gas plumes. Most of that work would be done in the months ahead at Paranal and at the W. M. Keck Observatory, atop Mauna Kea in Hawaii, together probably the two most cutting-edge centers for ground-based astronomy on Earth.

Mumma sat me down to lay out what was about to happen, as well as what they had found in the several nights before, when the "seeing"—the cloud, wind, and other atmospheric and deeper space conditions that affect astronomical data collection—was better than this night.

Mumma pulled up a map that showed where methane had been found and where the Mars Global Surveyor, a spacecraft that orbited Mars from 1997 to 2006, found the weak remnants of what had once been a strong magnetic field. The two distributions overlapped a lot. Mumma explained that a remnant magnetic field had been found in what is known to be the most unchanged, oldest Martian terrain. That meant, Mumma said, those areas had not been covered by the flooding lava of volcanoes or the transforming impacts of meteor strikes. (Mars is pockmarked with numerous craters, including the Hellas basin, the deepest one in the solar system.) Wherever the original Martian crust was modified significantly, the magnetic signature was gone. So usually was the methane, which generally was not found around volcanoes or lava flows, and definitely not in craters. The gas releases, then, were associated with an ancient Mars that was once well protected by a magnetic field, and by current estimations was most likely both wet and warm. In other words, a Mars that was at its most hospitable to life.

Water is the key to Mumma's methane work. Determining the origins of the methane alone would be impossible from Earth because there would be no telltale differences to measure. But methane on Mars along with water just might provide the necessary clues. So part of the search involves looking for water (as ice or vapor) and methane together. NASA has been trying to "follow the water" on Mars for decades, and now is quite convinced that it once flowed as a liquid on the surface and can be found in huge amounts as ice just below the surface. In 2008, the lander Phoenix found water ice inches below the surface on a polar plain, and in 2009 the Mars Reconnaissance Orbiter found ice much closer to the equator when it detected white, shiny material at the bottom of small craters recently created by meteorites. Many scientists believe the mystery of what happened to the abundant water of early Mars has largely been solved—some escaped into space but the rest is lying below the surface in huge reservoirs of ice.

All the recent discoveries about extremophiles, and especially those living in ice, raise the possibility that living creatures could remain in an arctic-like permafrost just below the Martian surface, one that changes from frozen to semiliquid with the seasons. NASA astrobiologist Richard Hoover recently found microbes deep in an Alaska tunnel at a level determined to be thirty-two thousand years old. They were inactive in the ground, but when warmed they came alive and even moved and, if they were methanogens, began to produce methane. Something similar could happen on Mars. Methane could also be stored in huge reservoirs produced long ago and now deep below the surface. It could escape when the warmer weather opens small pathways to the surface. And then there's ethane, a decomposition product of methane clearly associated with living things. Mumma is searching for that, too.

Mumma pointed on his screen to a large, ancient Martian volcano named Syrtis Major, traces of which spread 745 miles. His group had detected a methane plume around Syrtis Major, but it wasn't coming from the area around the volcano mouth. Instead, it came from an area to the other side that planetary geologists had determined to be quite unusual and perhaps the site of a deep underground collapse of what once had been a huge chamber of molten rock, or magma. Mumma was intrigued. "This collapse could create underground conduits, tunnels for the methane to escape. Or it could provide conduits that are colonized by the bugs that consume hydrogen gas and then produce the methane. At this point we don't know, but either would be consistent with the geology." Mumma's colleague Käufl couldn't help but suggest that the conduits could also be the fertile and protected home to large herds of farting cows.

For decades now, the conventional wisdom about Mars has been that no interaction exists between the subsurface and the planet's surface and atmosphere; that Mars once had volcanoes but magma is no longer coming from its depths; and that there are no hydrothermal vents or earthquakes on the planet, either. These destructive events are hugely important in terms of the origins and maintenance of life on Earth because they allow for essential elements and compounds to be cycled for use and reuse. Without Earthlike tectonic plates moving on the planet's surface to shake and heat things up, it's also hard to imagine how anything from below the parched surface could make it up and out without freezing when the temperatures average –81 degrees Fahrenheit. But a sense of how a release of gas or even water vapor could occur on Mars came to Mumma one day while he was driving up to visit relatives in Connecticut. It was a cold day, and it had recently snowed. Driving north, he passed through many cuts made through hills and mountains, and gradually the obvious pattern emerged: The north-facing sides of the road cuts were covered in ice, while the south-facing sides, which received more sunlight, were often dripping wet. In what passed for a near-eureka moment, Mumma blurted out to his wife, "That's Mars." The planet does warm up during summers and temperatures do travel above the freezing point in some areas, although not necessarily for long. But extreme forms of life don't necessarily need a long time to live and reproduce—like wildflowers sprouting in a desert after a rain.

More specifically, what he had in mind was information not contained in his initial published findings related to additional methane gas discovered in an area near Arsia Mons. That's a large (270 miles across) and ancient volcano in a region many miles from where they made their initial discoveries, but close to Olympus Mons, the thirteen-mile-high volcano and mountain that is the largest in the solar system, nearly three times taller than Mount Everest. Arsia Mons, he said, is home to the biggest mountain glacier on Mars (now buried, but as much as three miles deep when snow was falling in ancient times, climate modelers have concluded), and the area is filled with hundreds of miles of deep fractures in the ground, or "fossae," as such Martian features are called. It's also part of a line of volcanoes reminiscent of geology on Earth produced where continental plates meet and collide. Since volcanoes and plate tectonics play such an important role in enabling life on Earth, the possibility that similar dynamics were once at play on Mars was intriguing and suggestive. Methane, Mumma concluded, just might be seeping out of those kinds of cracks.

Even more intriguing was the area farther to the east, where there's a stress fracture in the surface, a deep gash that runs 500 miles long, 70 miles wide, and at points several miles deep. "Look here, what's the geology telling us?" he said, likening the big fractures to the rift valleys of eastern Africa, where the Earth was pulled apart by tension in the crust. "The net effect is to expose the cliff face, expose the layered strata of permafrost. Sunlight could certainly hit the edges, the faces of these scarps." That's where that model of the road cuts comes in, where the north-facing side has icicles and the south-facing side has water. Why wouldn't the same dynamic occur on Mars? "Crater walls, rock faces, they often show gullies, coming out from a layer below the surface—we don't know what, but stuff is coming. We think this is a possible mechanism for gases from below the surface to emerge when ice-clogged pores open in late spring and early summer."

At this point in our observing session the team had gotten Mars directly into its sights, which on the computer screen showed a bright ball between the slits created to focus the spectrometer. The image was blurry—not the kind of clear view you get from a powerful visible-light telescope—but you could make out some of the contours of the planet. Although the Mars on view was hardly spectacular, it was providing greater spectroscopic resolution of the planet than any image collected before because of the power of the telescope, the power of the spectrometer, and the increasingly refined use of a process called adaptive optics, which eliminates distortions created by the Earth's atmosphere. This is usually done by focusing on a guide star, but the team was delighted to find that for the first time they could achieve adaptive optics by locking on Mars itself.

Mumma takes a systematic approach to addressing the questions of the day: Is there definitely methane on Mars, where is it coming from, and what are its basic characteristics? Translated into scientific research, the overall goal of the campaign then becomes most pressingly to map Mars for methane and water, and to see where they coincide. "I want the planet, the whole planet, and in all seasons." Mumma asserts that good research practice makes it ultimately unimportant whether the Martian methane is produced geologically or biologically; the goal is simply to find which is the correct answer. Nonetheless, his working hypothesis appears to be that the methane is, or was, produced by organisms—that is to say, extraterrestrial life.

The goal of mapping for methane involves the age of the accompanying water vapor. It seems improbable, but the Paranal telescope and the CRIRES spectrometer can actually tell Mumma's team whether the Martian water vapor being detected is "new" (from the surface of the planet) or "old" (from its geological depths). If "old" water was convincingly detected in a methane plume, that would require a major reassessment of the long-held view that there is no direct interaction between the planet's lower depths and its surface and atmosphere. It would also significantly increase the chances that something alive is, or was, down there.

Pulling up slides on his computer with innumerable charts and graphs, Mumma explained that the water would be old if it had a lower percentage of deuterium, or "heavy water," in its H_2O vapor and new if it had a higher level in its H_2O vapor. Deuterium is an isotopic variation of hydrogen (with a proton and neutron in its nucleus, rather than just a proton) and that extra weight keeps it from sailing off into space as quickly as regular "light" Martian hydrogen does. The result of this process is that hydrogen in current-day H_2O (the kind found as ice at the Martian poles and circulating around the planet as vapor during some times of the year) has more deuterium and so is "heavier" than Martian water used to be. This dynamic is apparent in the famous Allan Hills 84001 Mars meteorite found in Antarctica. NASA scientists set off a huge controversy when they said the meteorite showed signs of ancient Martian life, but there was no real dispute about the determination that the meteorite was 4.5 billion years old and that its remnant H_2O was very light, with a ratio of deuterium to hydrogen at a very low 2. The Martian atmosphere today has an average deuterium to hydrogen ratio of about 5, meaning that much of the pure hydrogen has been lost. This is not just theory; scientists know and can measure these things. Although the difference is only one tiny particle in the atom's nucleus, the signature of heavy "deuterated" water on a spectrometer is easily distinguished from that of common "light" water.

And here's the clincher: Life generally prefers and produces lighter forms of its component chemicals; it's a pattern across all elements. So measuring which forms of hydrogen (or carbon) are present in Martian methane and water is a potentially big deal. If methane is found alongside "heavy" water, that means the gas is probably from near the surface; "light" water would mean it comes from deep below. Such are the clues, the inevitably indirect measurements, that will someday result in an announcement that Mars methane is or is not produced by living things—puzzle pieces understood only through rigorous science and no small amount of imagination and inspiration.

But the interplanetary forensics get ever more complicated. The spectrometer, which takes in light and other photons from the mirror of a telescope directed toward Mars, also takes in spectral information from the Earth's atmosphere, as well as light originating from the sun. One reason the formal unveiling of Mumma's methane-on-Mars paper took so long is that the team was creating models for measuring how much of the methane being detected was from Earth's atmosphere, how much from the sun, and how much actually from Mars. That's where Geronimo Villanueva comes in. An expert in applied physics, he worked out over five years the algorithms that allow the team to make these calculations. He also worked to make sure readings were not misconstrued because of surface weather and condensation on Mars, or because of a slew of other factors that could prejudice the results. Asked for a ballpark estimate of how many steps were involved in making his measurements—the building and calibrating of the instruments, the complex science of actual observing, and then the massaging and analyzing the data—he replied with a matter-of-fact geniality similar to Mumma's: "about as many as it takes to build a car."

Villanueva scrolled to a color-coded map that showed where on Mars they had found the methane. "We have this idea that Mars was very wet four billion years ago. If all the organics from that period and the water were stored in the subsurface and preserved there, then this is the place you can have a release," he said, pointing with an almost conspiratorial pleasure to an area in bright red. Villanueva, a native of Argentina, described their search as if the methane-water release was a plane ride away, rather than 155 million miles. "The moment you make that discovery of an active connection between the subsurface and the surface, then you open a Pandora's box of possibilities and processes happening. You can be talking about reservoirs of water, can be talking about biology and geology, a lot of things that are hard to think of in the hostile environment of the surface. The moment you see this release—boom—you have the discovery that Mars is wet in the subsurface." He's not talking about water ice; scientists already know that is there. He's talking about actual liquid water, kept warm by forces at work deeper into the planet. So Mumma and his team are not only exploring for methane, they're also trying to be the first to find signs of concentrations of liquid Martian water.

Mumma turned reflective about the challenge ahead. "In this kind of science, we can't really prove anything to be correct; we only can prove to be wrong. That's why we're always trying to confirm more. I learned pretty early on: Do not fall in love with your own interpretations and ideas. Be ready to accept a different view." When you've followed a path like Mumma's, when you've pushed back all your life against the world as it is presented, it is impossible to be content with what you think you know.

• • •

Now that the search for Martian life is focused on both methane and the microscopic creatures that we know can live in extreme environments, researchers have fanned out across the globe to identify, analyze, and better understand similar habitats. Pan Conrad, an astrobiologist for NASA and a co-investigator for the agency's landmark Mars Science Laboratory mission, has taken a lead in this far-flung fieldwork. Since 2006, she has mapped in intensive detail small sections of extreme places like the Mojave Desert and Mono Lake in California, as well as the McMurdo Dry Valleys of Antarctica and the Svalbard archipelago in northern, arctic Norway. She takes the temperatures of rocks and minerals, she scans for signs of static electricity and magnetic fields, she details the chemical makeup of whatever spare, frigid, or blasted terrain she is focused on, and she takes samples that will allow her to know what lives there and what organic compounds can be found. Death Valley does not leap to mind as an obvious place to study the conditions that allow for life, but actually few are better. In this spot, the lowest not covered by water in the Western Hemisphere, we might as well have been on Mars—which is exactly the point.

I joined Conrad at Badwater in California's Death Valley before dawn because she wanted to do some before-and-after measurements on the salt-mineral scramble: before dawn, when the ground retains its nighttime conditions, and again when the baking sun was high. She had defined a small area of jumble off a salt-slicked path and was into her third round of measurements when her voice rose in excitement: The static electricity meter was delivering a surprise. Laid on the crusty salt, the pocket-size meter was hardly moving. But when placed on the dun-colored toppings—mostly sulfur-based minerals—it shot up. She repeated the measurements several dozen times and found the same pattern repeated every time. "That," she said, "is just so cool. I have no idea what it means, but it's telling a story and I want to know it." We would return to the spot twice more that day, once in the afternoon heat and once in the dark of night.

Conrad is an expert, a pioneer really, in studying these extreme mini-habitats. She's a mineralogist by training, though she's also performed as a professional opera singer and was moviemaker James Cameron's companion for a trip to the Pacific floor in a Russian submersible. She began measuring and probing these miniature worlds after being selected in 2004 as one of several dozen investigators for a collection of instruments on NASA's next big mission to Mars, the Mars Science Laboratory. Three times larger than any previous Mars rover, the MSL is designed to travel as far as twelve miles from its landing site on its official quest to determine whether that small piece of Mars turf, selected with painstaking care, is or was ever habitable. That's a step short of the 1976 Viking goal of actually searching for Martian life, but because of the controversy and confusion over the Viking results NASA has never again flown a "life detection" mission to Mars. The agency still doesn't consider the time to be right, but the Mars Laboratory—scheduled to launch in late 2011 and arrive on Mars the next year—is a significant step in that direction.

The collection of instruments that Conrad and her colleagues are in charge of is called SAM, Sample Analysis at Mars. SAM's job is in many ways the most ambitious on the rover: to search for organic material, the kind of carbon-based compounds needed for life as we know it. Conrad's approach is based on this logic: MSL will be sending back the most detailed information ever about the chemical makeup of the rocks, minerals, and atmosphere of one promising destination on Mars, and if all goes well it will do that for several years. Because the location is certain to be extreme, like all of Mars as it's currently known, her team needs to know as much as possible about extreme environments on Earth so members can better understand the information that MSL will be sending back. She doesn't know what will or won't be helpful, but she wants to have an encyclopedia of extreme conditions data at her fingertips in case it becomes suddenly relevant. NASA's Mars mantra has long been "follow the water" as the surest path to habitable places and possibly life; Conrad's goal is to add other guideposts based on the presence, structure, and behavior of particular molecules, compounds, and minerals.

As she explained it, the actual job of looking for current or past extraterrestrial life is not what people imagine. Typically, extraterrestrials are envisioned as strange but visible, touchable creatures or vaguely human-looking aliens. But what Conrad, the MSL team, and future missions will be looking for is the presence, or former presence, of a life too small to see without a microscope—the kinds of microbes found in those South African mines or under Antarctic glaciers. As a result, they have to look at how that microbe may have changed the site's organics, minerals, and rocks, at the possible gases created by the current or past presence of a living bacteria-like creature, or at the chemical and electromagnetic landscape to see if it could conceivably support life. Physics, astronomy, meteorology, organic chemistry, spectrometry, the relatively new field of geomicrobiology—they all provide important tools in the forensics of extraterrestrial life. A lot of work for a seemingly limited return, but do remember what's at stake. There is a general scientific consensus that if life of any sort is found on another body in our solar system, and if that life has a detectably different origin than life on Earth, then all the calculations about life in the cosmos change dramatically. One genesis in a solar system and it could be a fluke. Two geneses and suddenly life becomes more of a feature than an anomaly; a cosmic commonplace. And if life is common elsewhere, then there's every reason to believe it has undergone evolution as on Earth, and could have become complex or even intelligent. One small microbe for Mars, one giant leap for life in the cosmos.

Intrigued by the different charges found in the brown crusts and the white carbonate "fluffies" all around them, Conrad decided to return later in evening with one of her big guns, a portable Raman spectrometer. A sophisticated (and, at twenty-three thousand dollars, expensive) piece of equipment, it can tell researchers in the field what molecules make up a particular rock or sediment or mineral they encounter. Used in mining and chemistry of all kinds, the Raman laser spectrometer (named after the man who theorized how it might work, C. V. Raman) has been adopted by astrobiology as an indispensable tool for analyzing other planets and celestial bodies, as well as their Earthly analogues. The portable spectrometer, called a "Rockhound," looks rather like a clunky but powerful ray gun, with a point-and-shoot laser beam that can harm the eyes or burn the skin. But the instrument loses its Star Wars menace when it's tethered to the Toughbook laptop computer it needs to perform.

Conrad wanted to come back at night because the Raman spectrometer works much better without light interference, and during the day Badwater was nothing but that. The moon was rising over the Amargosa Range on the eastern side of the valley just as the sun had begun to light the sky when we first arrived at the site sixteen hours before. We walked a ways on the salt pathway and, using her miner's light to scan the landscape, Conrad found another spot to analyze, this time pursuing "extreme science."

To help explain the differences in electrical charge, Conrad needed to know exactly what minerals and elements made up the white and brown deposits. She had already taken samples to analyze in her lab—which for twelve years was at NASA's Jet Propulsion Laboratory, but would soon be at the agency's Goddard Space Flight Center—but doing work in the field is part of her self-imposed training for the Mars mission. So she proceeded as if Death Valley were Mars, and set out to learn then and there what was putting out those very different electric charges and why.

Space missions are famous for their glitches, and we had one with the Toughbook and spectrometer. For more than an hour the computer refused to read what the Raman was picking up, and there were any number of reasons why. The instruments had been exposed to great heat in the car trunk, they were now being buffeted by wind in what amounted to an enormous salt bowl, and the temperatures were substantially below what they had been several hours before. Nothing so dramatic as conditions on Mars, but an extreme changeability nonetheless. The moon was fully overhead when finally the first squiggly lines of a spectral pattern showed up. Each peak on the graph is the signature spectral pattern of a vibrating bond between atoms in a mineral or compound; collected, the peaks reveal the identity of the ray-gunned materials. The graphs for the white valleys were, not surprisingly, very different from the brownish peaks. It was quite a sight. Conrad, seated on a tarp in a failed effort to get comfortable on the hard jumble below, the glow of the Toughbook, and another member of the team carefully moving the spectrometer ray gun from white to brown and back to white again. The moonglow kept away complete darkness, but it was well into the night and we were out on the Badwater salt with no other humans for miles around. It was noiseless, except for the wind. Mars in the day; Mars at night.

Death Valley is a popular site for Mars analogue research, but it's nothing compared to Svalbard, a region on the northern tip of Norway, well into the Arctic Circle. When the men and women who run Martian or lunar or other planetary experiments want to try out their equipment and learn the challenges, they now regularly go to Svalbard, an archipelago of islands best known for the town of Spitsbergen, their harsh beauty, and their three-thousand-plus polar bears. These animals are sufficiently fierce that when researchers (or any locals) go out of the towns, they are required to take a shotgun. It's a place that's often fogged in, where instruments can quickly die outside if not kept properly warm, where romances tend to flourish and breakdowns are not unheard-of, and where astrobiologists can do great research and can test their Mars or lunar rovers and other experimental equipment. And in the arctic summer, when the expeditions occur, it's light twenty-four hours a day.

The expeditions remain under the leadership of a Norwegian geologist named Hans Erik Amundsen, who did his doctoral research in Svalbard starting in 1997 and organized the first, almost impromptu international mission in 2003. The site has become valuable not only for testing instruments and training people on how to use them in Mars-like conditions, but also as a kind of scientific Outward Bound. "Nobody's allowed to work alone, and nobody goes ashore unless they've been trained on what to do if you encounter a polar bear. Everyone is in a group of maybe four to seven people with radio communication to the boat. Everyone has to be accounted for all the time, and at least one team member has a rifle and flare guns in case they come across a bear," says Amundsen. Many on the expeditions are high-powered scientists and engineers, and they often work by themselves and control their days. "They're alpha personalities, but all that has to be put aside and they have to meld into a team," is how Amundsen puts it.

The 2008 and 2009 expeditions focused on testing instruments for the Mars Science Laboratory scheduled to launch in 2011 and land on Mars in 2012. Although its mission is to determine "habitability" on Mars, scientists on the MSL team are convinced that the rover, the size of a Mini Cooper and weighing in at almost one ton, actually could detect life under certain circumstances. They say it, however, in something of a whisper. That's because the rover and all its instruments would have to be sterilized to a higher and far more expensive level if MSL were officially deemed a "life detection" mission. So being slightly less than that has its advantages.

Svalbard is a great test range for cutting-edge ideas, as well as new equipment. Mars scientists disagree about many things, but one that unifies most of them is the long-term goal of a "sample return" mission: sending a spaceship to Mars, collecting some especially promising rocks and minerals, and bringing them uncontaminated back to Earth via another spaceship. The challenges are enormous, but the United States does have an impressive track record on Mars—it's the only nation to ever land a spacecraft safely on the planet; the Russians tried many times unsuccessfully, as did the British with their Beagle spacecraft. But now, NASA and ESA have not only joined up for the ExoMars missions to measure trace gases like methane and then to land rovers on the planet in 2016 and 2018; they've also begun brainstorming and testing out ways to collect rock samples on Mars and bring them back to Earth. Svalbard has become a test site again for prototypes of the rover that will collect and then safely store the samples, called the Mars Astrobiology Explorer-Cacher, or MAX-C.

But the immediate project on everyone's mind is MSL, and the man in the spotlight for that is Paul Mahaffy, a two-time Svalbard expedition member and the principal investigator for the Sample Analysis at Mars, the many-faceted instrument on MSL designed to detect organic material and possibly life. A somewhat rumpled, low-key physical chemist, he will oversee a team of several dozen people who will conduct the most extensive and most significant investigation ever on the Martian surface for the kinds of organic, carbon-based materials that could lead to or be associated with life. Other instruments on MSL will definitively determine if minerals present were formed, as expected, in the presence of water, while another will shoot out an intense laser beam that will vaporize a small amount of nearby rock, allowing a spectrometer to then read much better what the rock is made of. Another instrument will be able to sniff for methane, which after Mumma's discovery has become a high priority. The reach of all these instruments will be greatly expanded by the ability of MSL to travel at least twelve miles during its two-year tour of Martian duty.

Mahaffy brought several off-the-shelf versions of SAM instruments with him to Svalbard, but the real SAM remained in a locked clean room at Goddard Space Flight Center. That's where I had seen it earlier, and it was quite a technological and aesthetic marvel.

In my short time with SAM, I was startled time and again by what it can do. It has, for instance, little heat chambers where rocks and sediments are baked into gases, and these can reach 1,000 degrees Celsius using but forty watts of electricity. Although it's tucked into the larger rover, SAM will be exposed to temperatures ranging from –40 to +40 degrees Celsius, sometimes in a matter of hours. It carries a "tunable laser spectrometer" that bounces received laser light some fifty times between two mirrors before it measures how the light is absorbed for gases like methane. The instrument can detect methane to a level of parts per billion. The maze of pipes and circuits and wires needed to run SAM so it can learn whether a particular piece of Martian rock has biologically produced molecules, or precursor molecules, or the extreme long shot of living molecules, is almost laughably complex. But the overall effect is of elegance, in part because the SAM containment box is plated in gold. Mahaffy said it's essential for controlling "thermal response."

Do scientists really think they'll find something—some sign of past or present Martian life? Probably nobody knows more about that question now than Steve Squyres, the Cornell University astronomer and planetary scientist who has led NASA's rover missions Spirit and Opportunity since they landed, packed into two large bouncing balls, on Mars in early 2004. These are the little rovers that could, the ones that were expected to pass out some ninety days into their mission, but instead were still going more than six years later. As they've wandered their little patches of Mars, they've collected more data about the planet than any other mission, and Squyres has been in the driver's seat the whole time. He's also now a Svalbard regular.

Results from the rovers, he said, "show more convincingly that Mars at some point in its past was a habitable world. You have to be careful, and I'm always reminded of the parable of the blind man and the elephant: We have two little, tiny spots on Mars we're looking at and we have to be careful about what we conclude. But the rovers are in very different places, and both show compelling evidence of near surface water, of interaction of that water with rocks and minerals, and in the case of Spirit site, you have hydrothermal activity—hot water and steam. These are the features that on Earth lead to local habitable niches."

It was quite definitive, and Squyres is hardly a starry-eyed newcomer. I met him at an astrobiology conference where he discussed and sought feedback about NASA's planetary sciences road map for the next ten years, an effort that he leads. I wanted to make sure I understood what he was saying, so I asked if the rovers have nailed that habitability question.

"I feel that they have," he replied, with a glint in his eye. "Yes."

6 THREE EUREKAS ON HOLD

You would think that a science that has fought so hard to be taken seriously would nourish a culture of dissent. But perhaps because of its urge for legitimacy, or because the discipline itself so often enters terra incognita, astrobiology has shown a consistent need to enforce a consensus. That is evident in the way it can treat those who diverge from the general view of what constitutes life on Mars and other celestial bodies.

Three reputable, diligent, and veteran researchers, for instance, are convinced that they have detected or even seen remains of life from Mars and from other celestial bodies. But not a single one can fill a hotel meeting room with their argument. Two are career NASA scientists who now carry the title of astrobiologist, and one was the creator and principal investigator for the main life-detection experiment sent to the surface of Mars for the two 1976 Viking missions. Together they have some seven decades of experience researching and analyzing experiments about extreme and extraterrestrial life, and two have been at center stage for some of the most important moments in NASA history.

Yet when the three took their places with others on a small side room dais at the Marriott Hotel in San Diego to address the topic of "Life in the Cosmos," few of the five thousand scientists attending the conference they were part of—hosted by the optics and imaging organization SPIE—were anywhere to be found. The Marriott's Marina Ballroom, Salon F, has seats for about 130 people, but on that summer night in 2009 only a quarter were filled. So it goes when the scientific community has concluded you're off base on a subject as charged as extraterrestrial life.

The first to unspool his findings was David McKay, the NASA researcher who introduced the world to the softball-sized meteorite from Mars that, for a short time in the mid-1990s, was hailed as containing strong evidence that life once existed on that planet. He and his team at the Johnson Space Center never claimed they had *proven* the rock, found in 1984 in the remote and meteorite-rich section of Antarctica called Allan Hills (and named ALH 84001), had been home to living microbes. Rather, they reported finding five distinctive characteristics of the rock, determined through various chemical analyses to be Martian and about 4.2 billion years old. Those characteristics are generally associated on Earth with microbial activity. McKay did report the possibility that the meteorite contained a Martian "microfossil"—the minute remains of the outer sheath of a bacterium—but his strongest results involved the presence of minerals and rock alternations that are considered signs that bacteria and other microbes were once at work eating the rock, transforming the rock, and depositing waste in the rock.

Not surprisingly, "proof of life on Mars" is the way the story played when it came out with a bang in 1996, and the "microfossil" was the star of the show. The discovery was featured in a major article in the journal *Science*, a full NASA press conference with two hundred reporters and cameramen present, and these words from President Bill Clinton: "Today Rock 84001 speaks to us across all those billions of years and millions of miles. It speaks of the possibility of life. If this discovery is confirmed, it will surely be one of the most stunning insights into our universe that science has ever uncovered." But in a foreshadowing of the bizarreness to come, the article was rushed into print because news about "life on Mars" was beginning to leak out. The source: a prostitute who was in the hire of Clinton political consultant Dick Morris. The president had apparently told Morris about the breakthrough, and Morris had told his companion, who then took the news to the tabloids.

Still, for a while McKay and his colleagues were on top of the scientific world, invited and feted everywhere. But like a pitcher whose no-hitter is spoiled in the ninth inning and who then loses the game, McKay quickly went from hero to goat. His team and their findings were subject to a fierce and wounding attack by other specialists in meteorites, geology, microbiology, and the study of ancient life-forms in rocks. That the Martian meteorite paper would inspire other scientists to study and criticize its methodologies and findings is hardly surprising—that's how science works. But that doesn't mean emotions and human defensiveness, offensiveness, and grandstanding weren't also at play. Attacking and defending the paper soon became a blood sport, an often brass-knuckled and highly personal struggle over the true contents and meaning of the meteorite. Suffice it to say that fourteen years after the paper was published, much of the scientific community has dismissed it, or at least concluded that it didn't offer the "proof" that it actually never purported to offer. McKay, who needed quadruple heart bypass surgery a year after the controversy exploded, has spent much of the intervening time responding to his critics, doing the experiments and reexamining the data in the hope that he can convince the world that the Allan Hills meteorite really did once house living Martian organisms.

Given this history, it was no surprise that McKay took shots that day at the Marriott at both his scientific critics and at a press corps that he said jumped on all the doubts raised about his work but seldom was interested when those critiques were found to be wanting. "What I would like to do today, if I can, is convince you that the Martian meteorite studies are very much alive, and furthermore that the evidence is becoming stronger all the time that Mars meteorites contain evidence for life on Mars," he began. "Now that sounds like an opinion, but I hope to reinforce that with some facts." What followed was a detailed and emphatic recounting of where things stood regarding both the original Mars meteorite and several others that McKay and his team have examined.

McKay's story goes like this, and is quite persuasive: His initial research led him to conclude that rock from Mars contained slightly magnetic grains or crystals that, on Earth, are often produced by bacteria that use and leave behind when they die signatures of the planet's magnetic field. The McKay team's original assertion that the meteorite held these magnetites was attacked as near ridiculous, since nobody had ever detected the presence or remnants of a magnetic field on or around Mars, either now or in the past. In addition, magnetites, like so many possible microscopic signs that living organisms once were present, can also be formed through nonbiological processes, and at least eight substantial papers have been written arguing for a range of origins that had nothing to do with life. The most common counterargument has been that the magnetites were formed when asteroids, perhaps the one that ejected the meteorite, hit Mars about fifteen million years ago and formed magnetites in the resulting shock and enormous heat.

But the McKay case was substantially strengthened in the late 1990s when the Mars Surveyor orbiter did detect remnant signs of an ancient magnetic field on the planet. It was a huge coup: McKay's prediction based on his early research—in this case, that Mars once had a magnetic field—was actually confirmed with observed data and measurements. Making predictions which are ultimately proved correct greatly strengthens any hypothesis. McKay then followed up with another rebuttal to his critics: His team published years of lab work that found the ALH 84001 magnetites were too pure and their concentration was too great to be explained by the contending "thermal shock" hypothesis—that they were formed in the scalding heat of a meteorite impact. In addition, a mineral needed to form magnetites nonbiologically was not present in the meteorite. Their conclusion: The magnetites were the mineral remains left by bacteria on Mars that, like some bacteria on Earth, contain magnetic crystals.

But McKay and colleagues Kathie Thomas-Keprta and Everett Gibson didn't get the rehabilitation they craved. Their foes kept insisting the original research was botched. For instance, Allan Treiman, a senior scientist at the Lunar and Planetary Institute, located in Houston, wrote in a 2009 paper that it was hard to disentangle the origins of ALH 84001 in part because the clues present have been "muddied by now-discredited claims for biological activity." Treiman later told me: "Would any reasonable person conclude that there was life on Mars based on the proportion of magnesium in submicron grains of magnetite?"

But McKay kept at it. His initial research had shown that the meteorites contained carbonates, which are minerals formed only in the presence of the liquid water needed for life. That finding faced the same hurdles as the finding of magnetites did: no confirmation of liquid water on or near the surface of Mars and the contention that the carbonates were formed in a superhot environment, where no life could possibly exist. But McKay's team was again helped by new discoveries on Mars, this time a fast-growing body of evidence that water had indeed once flowed on the Martian surface and that ice can still be found in substantial amounts just beneath the surface. McKay's colleagues had dated the carbonate globules at about 3.6 billion years old—when Mars was still potentially habitable—and subsequent tests have tended to support that conclusion. Considerable evidence now exists that Mars was much wetter and had a thicker atmosphere at that time. Lakes and even oceans likely existed. The idea that the carbonates were formed in a Martian cauldron is not heard much today.

McKay's microfossil was also controversial because no bacteria or other life-forms so small had ever been detected. McKay no longer argues as strenuously that the tiny microfossil is an important part of his case, because he is finding what he believes are much larger Martian microbes in other meteorites. Nonetheless, in the time since the paper first came out to such acclaim and criticism, many living organisms as small as the reported Allan Hills microfossil have been identified and categorized on Earth.

McKay's firm conclusion: The original meteorite did, indeed, show signs of ancient life. But—and potentially far more important—several other Martian meteorites he has since studied contain larger and more readily identified bacterial microfossils, and many are embedded deep in what he considers to be the undeniably Martian core of the rocks. McKay looks back on his time in the scientific trenches and wishes things had been different. "If we knew then what we know now, we would have had a stronger story that was much more resistant to criticism," he says. "Every bit of new Mars data is 'pro-life' in the sense of supporting the hypothesis that life has occurred on Mars. Every bit."

Where McKay and other students of meteorites are most vulnerable is in defending the rocks from the charge of Earthly contamination. The ALH 84001 and all other meteorites, many scientists argue, are corrupted with bacteria that move in as soon as they hit the planet and possibly even as they pass through the atmosphere. While some meteorites are collected soon after they fall, others remain undiscovered for eons; in the case of the ALH 84001 meteorite, the wait was about thirteen thousand years. That rock fell in the relative cleanliness of Antarctica, but other important samples fall in areas teeming with microbes, which can quickly penetrate the outer crust of the extraterrestrial rocks. Some pieces of the Murchison meteorite—the one that Glavin and Dworkin have used to explore chirality—landed on or near Australian farmland, other samples in the brush. Adding further to the threat of contamination, the Mars meteorites come filled with carbon compounds, which the microbes especially like.

McKay's answer to this criticism is that it is quite possible to tell the extraterrestrial microfossils from the terrestrial ones. But that view is highly debated, and involves some awkward, even painful history that still hangs over the astrobiology community.

It was a young but extraordinarily self-confident British microbiologist named Andrew Steele (or "Steelie," to those who know him) who first made contamination a major issue with Allan Hills 84001. He became involved with the Mars meteorite almost on a lark. Having just earned his doctorate at the University of Portsmouth, in southern England, he contacted McKay's NASA office soon after announcement of the discovery to say he wanted to get involved and asked for a small sample of the rock. Many others asked for, and received, similar samples, but few with so little experience and in such an informal way. Nonetheless, Steele brought a particular skill and technology to analyzing the sample and helped knock down one of the early criticisms of the McKay paper—that the "microfossil" was simply an artifact of the way McKay's team had prepared the thin slice of rock for examination. Based on his good work, Steele was invited to a NASA gathering convened in 1997 to discuss future research on Allan Hills 84001 and other Mars meteorites. That Houston meeting instead became an intellectual brawl, with scientists arguing endlessly about the initial article far more than planning any future research. At a key moment, Steele made some levelheaded and insightful comments that organizers of the meeting noticed. Then and there, he was asked to join McKay's research team. His soon-to-be wife was pregnant at the time in England, but Steele agreed to stay in Texas, and his future brother-in-law was hired as his doctoral student assistant. McKay and his team had already been working on the Martian meteorite for several years and had already published the *Science* paper before Steele showed up.

Perhaps it was his background as a microbiologist, or perhaps he just knew what to look for and how, but relatively soon after joining the team Steele came to McKay with some troubling news: A new sample of the meteorite that McKay had asked him to study certainly housed the remains of Earthly microbes on the outer crust and also in some cracks opening into the rock. As McKay tells it, the news was unsettling not only because it called into question what might have been "eating" the Mars meteorite and leaving those arguably telltale signs behind, but also because numerous teams of researchers had examined the rock earlier and not found the bugs. One of the teams that had searched for contamination but hadn't found it was McKay's—which was universally credited with being careful and rigorous in its research, even if many disagreed with its conclusions. The Steele finding led McKay to set up a Red or "Pro-Life" Team that would search for evidence to support their hypothesis and a Blue or "No-Life" Team that would poke holes in the research.

Steele was, not surprisingly, on the Blue Team, where he helped research and write some papers critical of the original McKay research. He spent only fourteen months with McKay before he left NASA for a dream job (for which he got a strong recommendation from McKay) at the renowned Carnegie Institution in Washington. He was on his way to becoming something of a science rock star—his good looks and long ponytail didn't hurt with that—and was named principal science investigator in 2006 for the long-term NASA-ESA Svalbard astrobiology project that tests robotic equipment designed to identify and assess signs of microbial life from afar. The Svalbard expedition has also played a role in the Mars meteorite debate because Steele and Treiman found carbonates with magnetites in them around an ancient volcano there. The two argued that the Svalbard magnetites were very similar to those found in the Mars meteorite, yet they were clearly not the product of biology. I went to visit Steele in Washington soon after the San Diego "Life in the Cosmos" event.

Steele shook his head and said he respected McKay, but that the samples were too contaminated to be of much use. "The bugs, they get onto and into the rocks in no time," he said. "And they can slip deep inside much faster than people imagine." (He once described the Murchison meteorite as contaminated "like a pig's ass.") What's more, Steele said, the microbes can quickly imprint their identities on the rocks through their activities. "You are what you eat," he said, meaning that if the Earthly microbes feast on the Martian organics in the meteorite, then the microbes will themselves take on the chemical and isotopic characteristics of the Martian rock. Microbes can flash fossilize and presto, you can have a microfossil that registers as extraterrestrial when it was alive on Earth not that long before. "I don't see any way around it—we can never really know what the meteorites might tell us about extraterrestrial biology because of all that terrestrial interference."

Steele said his experience led him to embrace a "null hypothesis" when it comes to extraterrestrial life. That means you assume that any apparent biosignatures in meteorites, detected by telescope spectroscopy, identified by missions to planets and moons, or from any other source, are contaminated, the product of nonbiological processes, or something left behind from the testing process. Only after all the possible nonbiological explanations have been thoroughly knocked down, he said, can biology come into the picture.

With good reason, Steele looks back on his time with McKay and working on the Mars meteorite as scientifically and professionally productive. McKay sees things rather differently. The day after his SPIE talk, I spoke with him about ALH 84001 science, its drama and personalities. McKay is a consummate gentleman, but he couldn't hold in his feeling about "Steelie." He volunteered that the young researcher was given way too much credit for work on the meteorite, something that he said "irritated me a bit." Steele "claims he was part of the team and I guess technically he was in the sense he looked at a few of the chips," he said. But the Englishman "played almost no part in anything we did, except to come in and find contamination on one chip of Allan Hills. . . ."

The way he tells it, McKay found the contamination. "I spotted the chip that was so contaminated with fungal material and asked him to identify it. I didn't think it was Mars at all, and he confirmed that it wasn't. But he then used that as a basis to say Allan Hills is entirely contaminated and you can't trust any identifications in it." McKay said that Steele did something similar with a sample of another Mars meteorite he was given to examine, one that fell in Nakhla, Egypt, in 1911, and contained a fungus that was clearly from Earth. Again, Steele concluded that the meteorite, which McKay sees as containing important corroborating evidence of Martian life, was too contaminated with Earthly microbes to be of any use. But McKay strongly disagrees. The fungus probably grew in Steele's lab, he says, and did not render the entire meteorite suspect. "I thought it was not very good science to make that kind of judgment," McKay volunteered.

So how do McKay and others determine whether a microfossil is from beyond Earth or is simply terrestrial contamination? The chemical composition of the microfossils is important. Newly fossilized bacteria, for instance, still have a lot of nitrogen in them, while ancient ones do not. Microfossils formed on Mars also would have different isotopic forms of carbon and some other elements. Nobody has a foolproof chemical method for telling what microfossils are terrestrial and which are extraterrestrial, so McKay has offered two other defenses of his conclusion: The microfossils in two additional Mars meteorites now being studied are encased in the mineral iddingsite in a way that could have occurred on Mars but not later on Earth. Iddingsite is formed only in water, he said, and water would not reach the inside of the meteorites once they hit Earth. Even more significant, the iddingsite near the outer zone of the meteorite had been altered by the crust that formed there as the red-hot meteorite sped into the earth's atmosphere. This heating occurred only when the meteorite fell toward Earth, so the iddingsite and whatever it enclosed had to be there prior to the journey. What he says is the inevitable conclusion: The microbes died and became encased within the iddingsite while on Mars.

Also, McKay points out that one of these Mars meteorites fell in Egypt and the other fell in Antarctica. Both have the same or similar microfossils, although the potentially contaminating microbes in the two locales where they fell are obviously very different. But after walking through these efforts to rule out contamination and more, he acknowledged that 100 percent proof will be difficult. His team is using the latest mass spectrometer analysis on the possibly fossilized microbes to try to confirm if they had once lived on Mars and if they contain organic carbon compounds. McKay would much rather work on carefully controlled Mars samples returned by robotic spacecraft missions, but that's at least ten years away and will cost billions of dollars. Meanwhile, nearly 100 kilograms of Mars meteorites are present in museums on Earth. Why, he asks, aren't more researchers devoting time and effort to these free samples from Mars while we wait for that illusive robotic sample return?

Richard Hoover is another astrobiology outlier, and he too took the stage in San Diego before that largely empty room in 2009. He is NASA's chief astrobiologist at the Marshall Space Flight Center in Huntsville, Alabama, and for more than twelve years he has used the agency's ultrahigh magnification electron microscopes to study some of the best-preserved and best-known carbonaceous meteorites to have fallen onto the planet. His conclusion is that some are home to large collections of fossilized remains of microbes that once lived on an asteroid, a comet, or some other place that wasn't Earth. In the mid-1990s, Hoover was the prime mover in setting up an astrobiology program within the SPIE, formerly known as the Society of Photo-Optical Instrumentation Engineers. He argued successfully that the search for extraterrestrial life will be an increasingly important issue for the men and women who make and use telescopes and other optical equipment. The group gave him its annual Gold Medal at the 2009 San Diego meeting in recognition of his work.

The Marshall Space Flight Center is known primarily for building rockets. It's where Wernher von Braun and his German rocket colleagues landed after World War II and later helped make the rockets that sent Apollo astronauts to the moon. Studying extraterrestrial life is not its strong suit. Yet Hoover, a longtime and respected microbiologist and expert in extremophiles and algae with silica shells called diatoms, was given pretty much free rein at Marshall to search for signs of life in meteorites after McKay's Mars meteorite was first introduced. Hoover does not study Mars meteorites, but rather some of the best-known rocks that have fallen to Earth from asteroids, comets, or perhaps other planets—legendary finds such as Murchison, Orgueil, Tagish Lake, or Allende. His specialty is carbonaceous meteorites, which are among the most ancient (some from the time the solar system formed), the least typical (less than 5 percent of meteorites), and the most interesting (because they contain the kinds of carbon compounds associated with life on Earth). They are, as a result, among the most precious and widely studied of meteorites.

To make his case, he puts on a slide show: First he shows images of a living cyanobacterium (a microbe sometimes called blue-green algae) and highlights some characteristics—long filaments made up of cells with small indentations where they meet. He also puts up an image of a cyanobacterium that appears to house a stack of dimes all just faintly connected. Then he puts up an image of a microfossil from the Murchison meteorite and it looks quite like the first, including the presence of trichomes (or strands of cells) in sheaths, and later he shows images from Murchison very similar to those stacks of dimes. It is quite a wild microbial zoo: Some of the microfossils are rod-shaped, some spherical, some sluglike, some almost slithery. The filaments often appear to be collections of once-alive cells and some clearly contain those telltale small indentations where the "cells" met. Hoover believes his menagerie constitutes "the unambiguous remains of extraterrestrial organisms."

At its most basic, identification of an extraterrestrial microfossil requires two findings: that the forms really represent the remains of a living organism rather than a never-alive but interesting formation in the rock, and that it comes in with the rock rather than infiltrating it once on Earth. To establish the first, Hoover showed samples from meteorites of forms that appeared to have cells and cell walls, that were in the process of splitting or otherwise reproducing, that were attached to the rock with what he identified as a primitive stalk called a basal heterocyst, and that had lived in what appeared to be colonies of microbes known to coexist on Earth. Hoover can identify sheaths of fossilized carbon filled in by the minerals of the extraterrestrial rock, and show with graphs that the presumed microfossils contain higher levels of the carbon required for life than the surrounding rocks. At the San Diego conference, he put up a hard-to-read image of hypermagnified rock with small rod-shaped grains and small filaments. "Now, these structures don't have very much morphology—it's hard to tell if it's really evidence of biological forms. But this isn't," he said with some delight as a much more clearly defined image appeared. "Here we see a sheath, and on the inside can see a chain of cells, rod-shaped cells, that make up the trichome of a cyanobacterium. We can see definite cross-wall constrictions." It was, he said, an obvious microfossil. Then came images with groups of the same. "We're not looking at single organisms, we're looking at an entire assemblage of organisms. There was a water body in the parent body of the meteorite, and huge masses of biology were growing at one point in time."

Looking at the bizarre elegance of the many forms, it was easy to imagine them once living on a large and watery meteorite, or hunkered down and barely surviving in the icy center of a careening comet, or perhaps on a distant moon. They didn't look like they could have been formed through crystallization, mineralization, or the other nonbiological forces that shape the interior of rocks. But as I knew from an earlier visit to his office in Huntsville, those images are maybe .004 inches long or less. They take on their life-size incarnation through the intervention of a very high-power electron microscope. The instrument magnifies so powerfully that it's sometimes impossible to go back and locate a previously discovered sample: The magnification is just too great, the size of the field too vast, and sometimes the geometric distortion is too confounding. Hoover says he routinely goes back to re-study individual microfossils; others say it is often impossible.

But Hoover had also measured the ratio of carbon to nitrogen in the microfossils and found that it was extremely high. Both elements are essential for life, and both are known to exist at particular levels in living organisms. When organisms die, their carbon is gradually transformed but remains identifiable while the nitrogen—needed to build life-critical amino acids, proteins, and DNA—is depleted slowly over millions of years. So as the age of the fossil increases, the ratio of carbon to nitrogen would also increase. His microfossils, Hoover has argued, have carbon-to-nitrogen ratios consistent with ancient life and inconsistent with modern microbial contamination.

As with McKay, contamination was a bedeviling issue for Hoover, so he has offered reasons to dismiss it as a problem. Many of his samples came from the Murchison meteorite, which has been a gold mine for researchers studying its amino acids and other organic compounds. As many as seventy amino acids have been identified in the meteorite, and some of them don't exist on Earth. If those organic compounds are extraterrestrial, how could it be that the microfossils are all contamination from modern microbes on Earth, Hoover asks? Murchison fell in the heat and scrub of Australia, and much of it was gathered within hours or days. So why, Hoover asks, has he found remains of organisms known to coexist in places like Antarctica and the arctic permafrost? How could they have contaminated Murchison? Perhaps most telling, he says, is that the microfossils have no nitrogen, an essential element for life. Nitrogen is lost from the remains of organisms over hundreds of thousands of years, which strongly suggests the microfossils are ancient. How, then, could they be the results of recent contamination?

If Hoover's analysis is right, why aren't his microfossils on the front page of every American newspaper? The answer involves some complicated history as well, because claims of microfossils in carbonaceous meteorites are not new. In 1961, at a meeting of the National Academy of Sciences, Fordham University organic chemist Bartholomew Nagy and microbiologist George Claus presented evidence for what they said were "organized elements" embedded in mineral grain of the Orgueil meteorite (which fell on southern France in 1864) and the Ivuna meteorite (which fell in Tanzania in 1938). Both were the rare carbonaceous meteorites and Orgueil in particular was so loosely formed that it would dissolve in water. Nagy and Claus didn't exactly say the "elements" were microfossils, but they provided supporting evidence that pointed in that direction. Other American and Russian researchers went further in asserting the meteorite contained remnants of extraterrestrial life. None other than Harold Urey, of Miller-Urey fame and a Nobel Prize laureate for other work, wrote in 1963 that "if found in a terrestrial object, some substances in meteorites would be indisputably regarded as biological."

A fierce and cutting debate ensued for a decade before a consensus developed that those "elements" were either contaminants from Earth or nonbiological but oddly shaped minerals that had tricked the investigators. It didn't help that one Orgueil sample from the Montauban natural history museum in France was found to have been intentionally or accidentally contaminated when someone drilled a hole and inserted coal fragments and seeds of a local reed, closed up the hole, and sealed it. As Hoover later wrote in a paper that began with a history of these events, "A few scientific works were subsequently published, but the suggestion of a hoax associated with the Orgueil meteorite won their debate and terminated serious scientific search for microfossils in meteorites for over three decades." Hoover told me that with this history in mind, he didn't want to take on American colleagues because he was afraid the association would eventually jeopardize their careers. He has instead turned to Russian scientists, collaborating with Alexi Rozanov, director of the Paleontological Institute of the Russian Academy of Sciences in Moscow, and working most directly at Marshall with Russian microbiologist Elena Pikuta.

Hoover began his microfossil work thinking he would reexamine some of the earlier meteorite samples for the "organized elements" described back in the 1960s, and he says he did confirm the initial detections—the ones that had been retracted by the original discoverers under professional fire. Yes, some of that meteorite was contaminated, Hoover says, but some of it was not and the two could be differentiated. He expanded his work to other carbonaceous meteorites and says he was completely surprised when he began to find the filaments and round coccoid shapes so common to the microbial world of Earthly bacteria. For some time he thought they were most likely modern contamination, but the chemical tests he did gradually convinced him they were ancient and extraterrestrial. Unlike David McKay, who did get worldwide attention followed by years of published critiques, Hoover's work has infrequently been cited by other researchers. He has his ardent supporters, but the circle is small. To better understand why, I sought out the head of the astrobiology program at the agency's Washington headquarters, Mary A. Voytek.

She began by praising McKay, who she said broke open the field of astrobiology, even if his conclusions remain controversial and not widely accepted. As she saw it, his team has continued answering critics with sophisticated lab work, and he continues to broaden his research and interpretations. "He has his work cut out for him—as he should, because of the importance of the discovery. I think he's really important to the field, which needs crusaders like him who have strong convictions and do good science." She paused and struggled to find ways to explain her different feelings about Richard Hoover.

"First off, it has to be said that the presence of microfossils is extremely difficult to demonstrate. You have to be very careful in asserting what you understand. You have to know so much about chemistry, biology, ecology to exclude other possibilities before you assert something could not be terrestrial and so has to be extraterrestrial."

Regarding Hoover's work, she argued that placing so much importance on the shapes of the objects is especially hazardous because circles, rods, and spirals are favored by nature, both in the biological and nonbiological worlds. What looks like a bacterial sheath, she said, could just as easily be a mineral formation. And then there's the contamination issue. Despite Hoover's arsenal of findings and extrapolations about why his samples are not and cannot be Earthly contamination, the field pretty much assumes that's what they are. Voytek dismissed the value of his carbon-to-nitrogen ratios as a sign of ancient life because the measure can be so misleading. Contamination scourge Steele told me: "I see the things Richard sees all the time. Everything tells me they entered the meteorite after it fell to Earth."

Voytek, who heads the NASA Astrobiology Program but doesn't supervise or fund Hoover's work, said the bar is extremely high for the kind of research Hoover has taken on, and that "in fairness to him" it's extremely difficult to do. She also said that as a microbiologist herself who has seen deeply held understandings in the field turn out to be incorrect, she cannot 100 percent rule out the possibility that Hoover *is* finding extraterrestrial microfossils. But "the community," she said, "has pretty much made up its mind that the work does not at this point support his conclusions."

Voytek had recently been up at Svalbard with the program that Steele helped set up, which has researchers use the instruments now available to astrobiology to assess from both afar and from close up whether rocks and other formations show signs of biomarkers. The work, she said, is excruciatingly complex. Her team's work on an object known to be a fossilized stromatolite, a primitive structure created by colonies of bacteria, illustrates the point. She said that using numerous instruments and working for days, the team was unable to conclusively prove it was in fact the structured remains of what once housed a colony of organisms—even though they already knew it was the case. She also suggested I look up a humbling 1982 paper in *Nature* by Jody Deming and John Baross, then at Johns Hopkins University and Oregon State University, that asserted active bacterial life had been found in a culture heated to 250 degrees centigrade. They had been researching the newly discovered worlds of "black smokers" at the bottom of the Pacific, hydrothermal vents where organisms lived in extreme heat. They brought samples of bacteria to the surface in pressurized containers and concluded they survived at those superhot temperatures in the lab.

Both scientists are very highly respected (and would later be selected to be members of the National Academy of Sciences) and their work was initially received with enthusiasm. But a researcher at the Scripps Institution of Oceanography, Jonathan Trent, was skeptical and so did some tests of his own in an effort to re-create the results. What he found was that all the measurements in the initial research were accurate in a sense, but that they were artifacts of the research itself rather than measurements of something that actually was happening. The sample had also picked up inadvertent contamination, which had not been sufficiently tested with controls. In other words, all the numbers were correct, but for reasons that had nothing to do with bacteria living in a medium heated well beyond the boiling point.

Still, research over the past decade into the worlds of extremophiles, microbes, and fossils has proven that what's true today often is overturned tomorrow, and what's rejected today may be accepted tomorrow. That is certainly what Gil Levin, the other prominent researcher in the world of astrobiology to cry "Eureka," hopes will happen. Levin has fought since 1976 to convince his colleagues and the public that the Labeled Release experiment on NASA's Mars Viking landers, a "life-detecting" effort that he designed, operated and analyzed, had succeeded. NASA does not dispute that his results were consistent with the agency's definition of what would constitute "life." But NASA and the space community also decided that this definition had been incorrect. The reaction captured by his Viking experiment was proof of biological activity, he says, but many other scientists say it was the result of an unsuspected geochemical reaction on the Martian surface.

The extraterrestrial guide star Carl Sagan had been a key member of the old Viking "biology" team, and at first he found Levin's results credible and compelling. Later he changed his mind, concluding they did not clear the high bar set by the standard he had done so much to popularize: "Extraordinary claims require extraordinary evidence." Levin has turned Sagan's admonishment on its head. "Something has happened in the last three decades. The claims are no longer extraordinary. My God, where do you go to find a place with no life? You can't do that on Earth. In our days as freshmen in biology class, we were taught that life was a thin, delicate film over the surface of the Earth. Well, that's now a lot of hullabaloo and we know life is everyplace—way above the Earth, on the Earth, way below the Earth, throughout the sea, under the sea, in rocks, down miles below . . . So it begins to seem as if life is an imperative. The big question is this: Is that imperative limited to Earth, or does it exist on Mars and elsewhere?" The answer, to his thinking, was obvious.

And it wasn't just limited, primitive, or unchanging Martian life that Levin imagined. He was amused, he said, by those who argued that if Martian life were present, it would exist only in moist subterranean pockets, and that it would not have evolved at all. "Darwin must be flipping over in his grave when he hears that. How can you have life anyplace and have it isolated for billions of years without adapting, without evolving, mutating too as it did on Earth? All of what we've learned about life on Earth, together with what we have learned about conditions on Mars, shows that the claim of the extraordinary is now quite ordinary. I think it would be quite extraordinary if Mars were *sterile.* After all, we know Mars and Earth have always been exchanging rocks, we also know those rocks, by experimental evidence, could contain microorganisms alive and deliverable from Earth to Mars in viable form. So it would be very strange if Mars was not populated from Earth, or Earth from Mars, or both from a third source. Our claims have become ordinary, while the evidence has become extraordinary."

First and foremost was the methane on Mars, strongly suggestive of current or past biology. Then the evidence for water, or shallow subsurface ice, on Mars today and strong evidence that it flowed freely and widely in the past. But even without flowing water, research into Earth's extremophiles has shown that bacteria can live in the liquid veins of ice and in mineral crystals. Now we know Mars had a magnetic field—necessary to hold a protecting atmosphere—and of course there are the meteorites that just might have Martian microfossils in them. And most edifying to Levin, the Viking test that had always been used to dismiss his finding—Viking failed to find any presumed necessary-for-life organic material on Mars—has itself taken a pummeling. Thirty years before, Levin had argued that Klaus Biemann's gas chromatograph mass spectrometer did not have sufficient power to pick up signs of low-level concentrations of organics. But Biemann was a respected MIT professor, and back then Levin had only recently earned a doctorate (in environmental engineering, from Johns Hopkins University). Biemann carried the day, with the help of lingering doubts about Levin's experiment, including the strength of the reaction he detected. But now an exact copy of Biemann's instrument had been tested in Antarctica and other harsh places where microbes are present but in limited supply, and it couldn't find the organic compounds and life that other instruments determined were there.

In 2007, it was Biemann who was writing an outraged article in the *Proceedings of the National Academy of Sciences* to defend himself against a strong critique of his Viking work by veteran researchers Rafael Navarro-González of the National Autonomous University of Mexico and Christopher McKay of NASA's Ames Research Center. Three years later, the two published a second paper asserting that, based on their Mars simulation work, Biemann's instrument most likely destroyed the organics present when it heated the soil to 930 degrees Fahrenheit. The researchers also wrote that when they heated organics and Mars-like Atacama Desert soil in the presence of the oxygen-chlorine compound perchlorate, now known to be present on Mars, they had found small amounts of two chlorine-based organics. The same compounds were detected in trace amounts in Biemann's Viking experiment but were dismissed as contamination from cleaning fluids. But the fact that Biemann's results may well have been wrong doesn't mean that Levin's are right. That is the conclusion that McKay and Navarro-González reached, to Levin's great frustration.

Many space scientists think Levin is something of a nuisance, someone who can't let go of a flawed experiment and result. Others see him as a pioneer and dedicated scientist who properly won't stand down. "Steelie," for instance, who did so much to undercut the work of McKay, says that Levin is one of his heroes. "He got his findings, he trusted the instruments, and he held his ground," he said. "I admire that." Scientists will eventually come forward with evidence of what they are convinced is, or was, extraterrestrial life, probably of the microscopic variety. How will we know when it's the real thing?

"Extraordinary claims require extraordinary evidence." Sagan's often-cited words have been used to bludgeon David McKay's Mars meteorite conclusions, Levin's Labeled Release results, Hoover's Murchison microfossils. The common refrain: How could they make such grand claims based on incomplete or controversial evidence? While Sagan's standard sounds right and may be entirely appropriate, the experience of at least one of these researchers shows just how difficult it will be to define and interpret. As McKay knows, but seldom says, one of the people given his initial Allan Hills 84001 paper to review for *Science* was none other than Carl

Sagan. He read it, no doubt had questions, criticisms, and suggestions, and ultimately recommended that it be published. He was terminally ill then, but still by all accounts alert and actively engaged in his work. Clearly, Sagan saw ALH 84001 as important—even "extraordinary"—science that was worthy of the attention it would soon receive. But fifteen years later, it remains just not extraordinary enough.

7 PLANET-HUNTING

You would think that searching for exoplanets many light-years away would require the newest, most sophisticated telescopes. But Paul Butler, one of the world's great planet hunters, has done some of his best work at an observatory formally dedicated in 1974 by Prince Charles in the Australian countryside. The telescope's mirror is relatively small by today's standards and the observatory has none of the power and élan of the Paranal facility used by Mike Mumma. The skies are far more likely to be clouded over and the telescope unusable at night than at other new observatories. It's also in kangaroo country, which Butler and many Australians see as a less than pleasant feature since the animals frequent the nighttime pathways around the campus: A full-force kangaroo kick can kill a man.

But Butler loves the Anglo-Australian Observatory (AAO), twenty miles outside the two-street town of Coonabarabran in New South Wales and next door to the dramatic and ancient volcano remains of Warrumbungle National Park. Some of its perceived weaknesses, he says, are invaluable strengths. Yes, finding extrasolar planets is hard, but it's well within the capacity of the four-meter Anglo-Australian Telescope (AAT) glass. And because the observatory is no longer cutting-edge, and is in a distant location that American and European astronomers are not keen to put in the time to use, Butler has what he needs most—lots and lots of nights on the telescope. He spends three months a year at the AAO. That means he can study a star for hours, even days or weeks if he suspects there's a planet circling it. Since he began coming to Coonabarabran, the self-proclaimed

"Astronomy Capital of Australia," Butler and the team he leads have identified and to some extent characterized 40 exoplanets. Worldwide, astronomers have found more than 500 exoplanets, and Butler has been in some way involved in about half those discoveries. He plans to planet-hunt for years to come, but now he has a new focus: the makeup or "architecture" of solar systems. That's because discovering where exoplanets are in relation to their suns and companion planets is essential to determining if they're habitable and could ever be home to extraterrestrial life.

The son of a Los Angeles policeman, Paul Butler is a tall, bearded man who loves jazz and wears long pants only when it's below freezing. He's been looking for exoplanets since the late 1980s, more than a half decade before a Swiss team announced the first detection, which soon after was confirmed by Butler and his colleagues. He's away from his wife, his home, and his office at the Carnegie Institution in Washington more than half the year at observatories in Chile, Hawaii, and in the land of kangaroos (or, as he terms them, giant-tailed rats). He calls the AAT control area his living room because he's been there so much and he feels that comfortable in it. It helps that the telescope is also just through a blackout door from the control room; at more sophisticated and more highly elevated observatories like the W. M. Keck in Hawaii, the control room is a two-hour ride down the mountain. The nighttime world of star- and planet-gazing exerts an almost gravitational pull on those captured by it, an endless desire to know more about the mysterious yet increasingly knowable vastness in which we live. "On clear nights, there's absolutely nowhere I'd rather be," Butler says.

But our first night at the telescope was cloud covered—actually, was socked in and packing knock-you-down wind from a cyclone on the eastern coast—so instead of observing, we talked. Butler was eager to explain exactly why the AAT remains such a godsend even though significantly more sophisticated telescopes are available, and his discoveries about the planets orbiting the star called 61 Virginis were exhibit A. The research was published in 2009 and represents the teasing out of one of the first three-planet solar systems orbiting a sunlike star.

"Look at this curve," he said, pointing at a computer screen full of initially indecipherable but nonetheless elegant graphs, the kind that allow him and other planet hunters to determine that a distant planet is present. Specifically, he motioned to the chart labeled "61 Vir," which happens to be one of the closest bright stars to Earth and one that is visible to the naked eye. "We'd been observing that star for years, and now we were seeing something. But we had to pull out the signal, and it was very complicated." Astronomers will one day be able to routinely see or "image" distant planets using considerably more sophisticated telescopes than those available today, but for now most planet hunting involves indirect measurements of the effects of extrasolar planets on their suns.

First Butler and his team found signs of a planet the size of five Earths orbiting 61 Vir in a breakneck four days. "But after a while the pattern changed—something that doesn't happen unless there's a good reason. We suspected there was another planet, and that one turned out to be Neptune-sized with a thirty-eight-day orbit." But still something was off, and Butler wouldn't know what it was until he and his team observed for several more weeks. What they ultimately found was the signature of not one planet but of three: one orbiting closely, one at 38 days to circle its sun, and then another at 125 days. This only became clear because Butler had weeks of time on the telescope—forty-seven straight nights, a run that would be impossible at the bigger and more sophisticated observatories in Chile and Hawaii. And all those nights of observing allowed his team to put enough dots on its chart so it could read the complicated message being sent by the planets.

When astronomers began detecting exoplanets in the mid-1990s, it was very big news that landed a story about Butler and his colleague, Geoff Marcy, on the cover of *Time* magazine. The implications were as exciting as the discoveries themselves: If planets were found to orbit tens of billions of stars in the Milky Way alone, then it seemed entirely plausible that some contained liquid water, nutrients including carbon, and an atmosphere to keep out the most damaging cosmic and solar rays. In other words, the basic conditions for life as we know it. And who knows, some of those planets could well be home to life as we don't know it, based on different chemicals and conditions. Suddenly the prospects for extraterrestrial existence increased dramatically. It was no coincidence that the emergence of astrobiology as a respected and, soon after, a hot field of research occurred in the mid- and late 1990s—right as it became clear that the discovery of the first extrasolar planet, 51 Pegasi, would be followed by many more. NASA started its formal Astrobiology Program in 1998 with these exoplanet discoveries, as well as that controversial announcement of signatures of life in a Martian meteorite, very much in mind.

But as more planets were found, it became clear that many, and probably most, were strikingly different than what almost all astronomers and planetary scientists expected, what Butler calls the "Everything You Know Is Wrong" phase of extrasolar planet research, borrowing from the Firesign Theatre comedy team of the 1970s and one of its iconic acts. The consensus of the astronomy community had been that distant solar systems would be similar to ours, that the known physics and dynamics of star and solar system creation required a certain kind of arrangement. Yet huge Jupiter-like planets were discovered revolving extraordinarily close to their suns. In fact, planets with highly eccentric orbits were found to be far more common than the near-circular orbits of planets in our solar system. Even more unexpectedly, solar systems were found that were somewhat like ours but with seemingly impossible variations—for instance, with a circular-orbiting Jupiter in what is considered the roughly "right place" in relation to its sun, along with an eccentrically orbiting and even larger Jupiter in the inner solar system region where rocky planets are supposed to live. As it turns out, Butler said, having our solar system as a model "can be worse than having a sample of zero because it leads you down one road and you don't imagine the others." But because of research like Butler's, the field of planet hunting has abandoned its previous assumptions and now is working hard to make sense of the new reality that solar systems structured like our own are a distinct minority.

None of these discoveries were, or are, particularly good in terms of the search for extraterrestrial life. But they're not the final story at all; rather, they're scientific waystations along the path to detecting the Earth-like planets virtually all astronomers believe exist, and an introduction to the kinds of solar systems to avoid if finding habitable zones and distant biology is your goal.

For instance, a consensus exists within the astronomy community that to have any chance of supporting life, a solar system needs a huge Jupiter- or Saturn-sized planet (300 and 100 times more massive than Earth, respectively) in roughly the locations where they sit in our solar system. That's because the gravitational force of the giant planets serves to pull in and destroy asteroids and other celestial bodies that might otherwise head into the "habitable" zone and smash the small rocky planets to bits. This is why in astronomical circles Jupiter is often called our protective "big brother" or "big bouncer." But if Jupiters and Saturns in many other solar systems are close in to their suns, or otherwise in what is considered the wrong place, then they can offer no protection at all. The question of eccentric orbits is perhaps even more unsettling. A planet that swings very close to and then very far from its sun will almost assuredly experience temperature swings too extreme to support life. We know that living things can exist in very hot and very cold environments, but the same organisms probably can't exist in both. In addition, the gravitational force of a large planet with a strongly eccentric orbit would most likely kick any smaller planets out of their solar system and into space. Nothing personal—it's just gravitational physics at work. "Solar systems with really eccentric orbits are about the worst place to look for life," Butler says.

So while Butler and his colleagues continue their two-decade search to detect and characterize extrasolar planets, the new and most important questions in the field have changed and become quite a bit more complicated and ambitious. With Miles Davis and John Coltrane as the backdrop to his thinking, Butler described the two goals that he hoped to help achieve before his time as a peripatetic, globe-hopping astronomer comes to an end. Like his stargazing colleagues everywhere, Butler speaks in a language that can often seem mysterious and impregnable—throwing out references to laws of physics, categories of stars and planets, and modes of measurement that are foreign to the uninitiated. The concepts behind them, however, usually make a pleasing, even elegant kind of sense:

"Overall, what we're trying to find is solar system analogues because we'd like to know how common or rare the architecture of our solar system is. What are the systems that have Jupiters and Saturns beyond four or five AUs [astronomical units, or the distance from the sun to the Earth]? What are the systems in circular orbits? Those are the signposts for us—systems with noneccentric orbits and with big brothers to shield the smaller inner planets. That's where you'll find habitable zones with the potential for life. When we find them we want to go back and stare at them hard and look for the Earths and other inner planets [that] should be there." He said it would probably take another ten years of planet hunting to get a good representative census of solar system architectures, and that the percentage of systems like ours might be as low as 5 to 15 percent—a perhaps disappointing number until you recall that there are trillions of trillions of stars out there.

"Second, we need to know about habitable planets, how common or rare they might be. Right now our best guess is that rocky planets like Earth and Mars in zones where life could theoretically exist are present in most solar systems, but we really don't know and could be all wrong. Finding them will be hard because of the ways we look for planets, and so it may seem that any clear understanding is way off." But people are making progress, he said, and right now his team can find planets only five times the size of Earth in habitable zones—that is, positions in relation to their suns where they are likely to be rocky planets with liquid water—around M dwarf stars. M dwarf stars, or red dwarfs, are the most prevalent in the sky. They are about half the size of our sun and produce far less energy, making it theoretically possible for habitable planets to orbit in close, where current planet-hunting technology can better detect them.

The technology and know-how for finding planets has exploded in the past decade. It's growing at an ever-faster pace alongside, and to some extent because of, the exponentially increasing speed and power of computers, and that has instilled a broad confidence in the astronomy and astrobiology communities that the future will be full of discoveries, including planets the size and consistency of Earth. It also doesn't hurt that while the number of planet hunters could be counted on one hand when Butler started, he estimates there now are about a thousand. "We're just on the hairy edge of this," he said regarding the discovery of those Earth-sized stars. "I'm convinced it's doable, and actually is beginning to get done by my group and two others"—the University of California at Berkeley group that Butler used to be part of and the Geneva Extrasolar Planet Search in Switzerland. This is not necessarily a majority view, because the technical challenges of finding Earth-sized planets remain daunting, and the best approaches now possible cannot definitively locate a planet the size of our ball of rock. But Butler says his team could find a planet the actual size of Earth right now if they had six full months of consistent observing time at a major telescope—a coup, however, that even he can't pull off.

Although the Hubble telescope has "imaged" a handful of distant planets from its perch above the atmosphere, virtually all the extrasolar planets identified have been discovered by astronomers and astrophysicists using ground-based telescopes. The process by which Butler and his colleagues find their extrasolar planets is both oddly simple and confoundingly complex: A big telescope (usually 4 to 10 meters in diameter) takes in billions of photons coming in at the speed of light from a selected star, bounces the light waves through a maze of mirrors that shape them in the desired way, and focuses the light through a narrow slit into a spectrometer, where several prisms and other optics divide the light into its spectral parts. That spectrum is then photographed at extremely high speed (by a camera cooled to –300 degrees Fahrenheit) and the image is embedded into a high-end CCD chip not dissimilar from the one that makes your digital camera work. The result is the production of a single number, which then gets massaged a little further and ultimately placed on a graph, which captures the "wobble" of a sun caused by an exoplanet. At the AAT, Butler likes to say, 4 billion photons captured and worked over by his team of Americans and Australians become one data point. The process is then repeated scores, even hundreds of times to learn about the neighborhood of a single star.

The AAT was built when thick steel and heavy concrete were still in astronomical vogue, so it looks like a huge but comprehensible machine. The supposedly "simple" part of planet hunting is how and why those points on a graph can tell Butler whether a planet is orbiting the star he's observing. The key to Butler's team's technique is harnessing the Doppler shift, a phenomenon in physics that can be used to measure the speeding up or slowing down of just about everything. The Doppler shift was discovered and described by Austrian Christian Doppler in 1842. He, like many others, had been intrigued by the sound of a train whistle as it approached and then departed from a station it was passing by. The whistle, always the same in volume, nonetheless sounded different as the train approached, as it sped by, and as it left the area. That difference, Doppler concluded, was the result of a change in the frequency of the wavelengths, or of their pitch.

As scientists noted in the ensuing century and a half, the same effect occurred in waves of all sorts—with light, with X-rays, with radar and radio waves, and so on across the electromagnetic spectrum. Invariably, when a wave approached or receded from the observer's line of sound or sight, the perceived frequency changed in a measurable way. The trick was finding ways to capture that information and measure changes in wavelengths (or frequency) in relation to an observer, whether at a nearby train station, along a highway where police officers use radar guns based on the Doppler shift to identify speeders, or at observatories where astronomers were looking for ways to detect the minuscule changes in photon wavelengths that would be associated with distant stars that had planets orbiting around them.

Classical physics tells us that the gravitational force or pull of any and all extrasolar planets would have an effect on the suns they orbit. A star with a planet around it would wobble ever so slightly in its own orbit because of the planet's pull. A star without exoplanets wouldn't wobble at all. The gravitational pull of Jupiter, for instance, makes our sun wobble around in a circle at a speed of almost forty feet per second. Astronomers have used their knowledge of Doppler shifts for more than a century to measure the movement of stars, but it wasn't until the 1980s that scientists—and especially a team from Canada that came very close to finding the first extrasolar planet—actively began using it for planet hunting, a technique called "precision Doppler velocity." In effect, they captured and analyzed all those photons to determine the speed at which a star was moving toward or away from the Earth. When those single measurements were charted over a long period of time, they showed a straight line (a star with no planet-induced gravity tugging it one way or another) or an undulating line that told astronomers that the star was being tugged ever so slightly.

What Butler and his colleagues measure is akin to the speeding up or slowing down of a person walking, in a figurative sense, on the stars, which are roiling, gurgling balls of gas a million miles in diameter. If the speed of the star moving toward the Earth or away from it changes by as little as one meter per second due to the gravitational pull of an orbiting planet, Butler and his colleagues can detect this. That rate, one meter per second, is strolling speed, a not even particularly brisk walk, yet the planet hunters have used changes of that size embedded in light traveling to Earth from light-years away to actually prove that the planets are present. Not only that, they can determine as well the shape of the planet's orbit—circular or eccentric—from the same information.

At the AAT control room, all this translates into a pretty low-key affair that can involve as few as two people. A technical operator opens and closes the dome that protects the telescope, and solves a problem if one comes up; an astronomer who determines which stars to observe watches over the data as it comes in by computer, and decides to move on to another target when the "seeing"—the telescope's ability to take in enough photons from a star—starts to decline. The actual analyzing of information takes place in computers back in Washington, D.C., although Butler can connect to them from Australia and has actually found several planets while sitting in his AAT living room.

Once the big dome is opened and the telescope is moved into place via gears that grind with a slow, hollow moan, computer screens light up and numbers and graphs appear. But the most exciting action is at the technician's console, where light from the star itself pours into a slit displayed on the screen and then is further directed between horizontal bars about a half-inch apart. The concentrated starlight jumps and dances between and sometimes outside the bars, movement you might expect from a star with a dynamic and always churning surface. But that's not really what's on display. The movement tracks how many of the star's photons are making it to the telescope and through its spectrometer, and any dancing outside the bars is considered bad. All this can now be corrected by computer, but not that long ago astronomy grad students would spend long hours with a joystick working to make sure the starlight remained within the narrow bars.

The weather was turning bad again and so Butler decreed it was time to drive down the hill to Coonabarabran, a three-pub town of some 2,500 people that works hard to both maintain a connection with the observatory and to benefit from its proximity. We headed for the Imperial Hotel pub, an old-school, dark-wood and tinted-glass establishment in just about every way, except for the wall-to-wall carpeting in azure blue with a star and planet motif, and a darts and video room called "Galaxy Games." The technology, pace, and culture of both the observatory and Coona were definitely 1970s and '80s—with a lingering dose of mid-twentieth-century small-town America, or its Australian equivalent. That made it a fitting place to pick up the story of how a kid growing up in a Los Angeles police family came to be one of the world's premier planet hunters.

Butler was born in 1960 and became interested in astronomy as a teenager. As is often the case with similarly minded kids, he wanted to build a telescope. In his time off duty, his father, a thirty-year veteran of the force, used two-inch plumbing pipes to fashion the steel struts that supported the instrument, and accompanied him to an amateur telescope contest at California's Tehachapi Mountain. Butler developed his interest in astronomy the old-fashioned way—in a library. He sampled books from aisles of all sorts until he reached the astronomy section. It wasn't so much the science that first reeled him in, but rather the brilliance and inventiveness of the earliest astronomers like Tycho Brahe and Johannes Kepler. They're the ones who pretty much established astronomy as a science, where the goal became taking precise measurements, determining the level of error inherent in the calculations, and beginning the process of demystifying the cosmos. But just as eagerly and importantly, Butler read about the terrible fate of Giordano Bruno, a lapsed Italian monk and freethinker who, in the late 1500s, advanced the idea that planets orbited a universe full of moving distant stars—an idea that challenged conventional and ecclesiastic belief at the time. Bruno wasn't the only one proposing such things and he did not have the scientific instincts or knowledge of a Brahe, a Kepler, or a Galileo. But he definitely got on the wrong side of the Inquisition and was burned at the stake in Rome in 1600. One of his several heresies was the belief in and advocacy of the notion of "multiple worlds."

Butler left home for San Francisco State University—a good school but hardly an academic powerhouse—and met a junior professor named Geoff Marcy. The teacher was getting nowhere in his work on the magnetic fields of stars while Butler was finishing up a bachelor's degree in chemistry and a master's in astrophysics. Marcy, as voluble as Butler was then reserved, recalls an epiphany in the shower, when he realized that to succeed in his chosen field, he had to address the kind of questions that he had cared about as a child—including whether planets orbited other suns. Butler's background in chemistry came in handy because to achieve the level of precision needed to identify extrasolar planets via the gravitational wobble of their suns, they had to find a precise, safe, and stable element to serve as an absorption cell as the light entered the spectrometer, a cylinder filled with a gas that serves as a guidepost or measuring rod of the incoming spectrum. Previously, the Canadian team had used hydrogen fluoride as their standard, but the compound was both insufficiently precise and extremely toxic, requiring extreme and time-consuming care since the smallest exposure could be harmful or even fatal. Butler came up with the idea of using an absorption cell filled with iodine, a breakthrough that allowed the duo to move ahead more quickly with the task of increasing precision while avoiding the potentially lethal dangers of the hydrogen fluoride. Butler's iodine absorption cell became and remains the standard in the field.

What began as a collaboration became an obsession as Butler and Marcy tried to reach a level of precision in detecting those Doppler shifts that would finally allow for a planetary detection. When they started, they could detect motion if it reached 300 meters per second. But to be of any use, they had to bring that number down to a mere 3 meters. Today the goal is to get dependable measurements when the wobble consists of motion under 1 meter per second. "When we started in the eighties, we weren't thinking of aliens and life—we just wanted to find a damn planet. People have been thinking about extra solars since 1600, you know, when Bruno was burned at the stake for that. I wanted to help solve a scientific problem that could get people that upset, that still had such a huge power to affect people." He and Marcy were two very smart kids who, despite their long distance from the Ivy Leagues, were determined to make a name for themselves. The process of refining their technique took eight years, saw them secretly commandeering time on colleagues' unused computers at night to process their data, and led them down blind alleys that ate up years at a time. "Basically, we had to swing for the fences or we would never get anywhere," he says. Many aspire to be home run hitters but few succeed quite so spectacularly. As the discoveries rolled in, the two received about every prestigious award handed out in their field, including one from the National Academy of Sciences.

Like many important players in astrobiology, Butler didn't begin with any particular interest in extraterrestrial life, or any quixotic drive to find it. But, he says, "no question, we're not just totally disinterested robot scientists—we're humans, right, and we want to know things that reflect on us, that tell us about what we are and where we came from. So we're interested in earths, we're interested in solar systems. These other planets that we've found have been shocking and amazing, but they bring even more to the center that 'Just who the hell are we?' question. Where did we come from and how common are we? We just keep on coming back to looking for our kinds of earths and our kind of solar systems, and I imagine we'll continue until we find them."

As I learned ten months later, Butler wasn't just speculating about finding "our kinds of earths." Seated halfway around the world beside excited and smiling officials of the National Science Foundation outside of Washington, Butler and his research partner of fifteen years, astronomer Steven Vogt of the University of California at Santa Cruz, announced they had detected a planet only three to four times larger than Earth in an apparently habitable zone. Based on eleven years of data collected at the Keck Observatory, they concluded that planet Gliese 581G, an astronomically close 20 light-years (or 117 trillion miles) from our solar system, had the mass to hold an atmosphere and was at a proper distance from its sun to hold liquid water. They both said it was plausible that life could exist on the planet but also that Gliese 581G is definitely no Earth clone, since it always faces its sun just as one side of the moon always faces Earth. Still, they said the regions where full-time sun shaded into full-time dark were large enough to support living systems, if a variety of other conditions were met. "This is our first Goldilocks planet—just the right size and the right distance from its sun," said an ebullient Butler, who had reluctantly broken his no-long-pants rule in warm weather for the event.

The announcement was met with some initial skepticism in the exoplanet community, and Swiss planet hunters exploring the same solar system said they had searched their data and had found nothing like Gliese 581G in a potentially habitable zone. Butler replied that the Swiss had substantially less data to make their assessment, though he acknowledged that he and Vogt had published with a slightly lower confidence level than most of their previous findings. The discovery was at the "raggedy edge" of their planet-finding abilities, he said, and does need to be confirmed by others.

A paper two weeks later in the journal *Science* by Butler's former partner Geoffrey Marcy made clear that regardless of whether Gliese 581G is a habitable Earth-sized planet, there are certain to be many, many other candidates out there. In fact, based on observations at the same Keck Observatory, Marcy and his University of California at Berkeley partner Andrew Howard concluded that Earth-sized planets are absolutely common in the universe. So common, he said, that tens of billions are likely to exist in the Milky Way alone. Not all, and probably not most, Earth-sized planets are actually like Earth. But it is now substantially easier to imagine a galaxy and a universe with rocky planets very similar to ours in all the important ways. "A threshold has been crossed," Butler declared.

As coincidence would have it, one of the pioneers in the field of understanding the atmospheres and mantles of faraway exoplanets used to have an office at the Carnegie Institution just down the hall from Butler. She is Sara Seager, now at MIT. By understanding how to collect enough data to seriously analyze those atmospheres, Seager and others maintain, we will someday be able to detect molecules and compounds that are intimately related to life. Oxygen, ozone, and complex carbon compounds like methane would, if found, strongly suggest that some kind of life-form was on the planet and producing those molecules as a by-product of its existence. The techniques being developed now would not provide a 100 percent certain detection of extrasolar life, but would be able to predict to within the 95 percent range that some life-form was responsible for a biologically produced gas if found.

Seager works on the data collected by others, and most centrally she's a theorist. Her theoretical understandings of extrasolar planets, for instance, helped lead a colleague to prove that a previously unsuccessful but important method to detect distant planets could and did work. Using the

Hubble Space Telescope, David Charbonneau was the first to find a planet transiting its sun by measuring the small drop in light that accompanied that passage. Harvard-Smithsonian Center for Astrophysics astronomer David Latham, whom Seager repeatedly urged to explore the "transiting" technique, confirms that Seager did indeed play a "catalytic role" in the subsequent discovery because he passed on the prediction and a transiting candidate to Charbonneau, who was also one of his students. In his 2003 paper reporting on the discovery of sodium in the atmosphere of an exoplanet 150 light-years away, he specifically wrote that another prediction made by Seager was central to his successful effort.

While Seager's work can be highly technical and demanding, what makes her unique is her desire and ability to make the world of planet and atmosphere hunting accessible to a general audience. That and her cut-to-the-chase willingness to say she's in the exoplanet fray because she wants to find the life she's confident is out there—to be part of science and exploration that will someday significantly change our understanding of the Earth, human life, and the universe. I got an introduction to her effect on people at a most unlikely venue: an outdoor trattoria in one of the old quarters of Rome. She was in town to deliver a talk at a conference on astrobiology called by the Vatican's Pontifical Academy of Sciences, and we met on a day off from the proceedings.

We were seated at the very edge of the covered section of the restaurant, and the tables were set close together and filled. Seager spoke with her customary energy and command about topics such as how to find rocky, Earthlike planets thousands of light-years away, about why learning the chemical makeup of exoplanet atmospheres is a key to finding distant extraterrestrial life, and about the sumptuous but awkward pleasures of being put up on Vatican grounds in what had been the pope's villa. She grew especially animated about the possibilities of launching an "occulter"—a flat, football-field-sized, petal-shaped sunshade that Seager (and others) hope will one day be sent into deep space and aligned with a similarly stationary but space-based telescope. It's essential because an inevitable and unchanging problem for all exoplanet work is that stars are vastly brighter than the relatively minuscule planets orbiting them. The challenge, Seager explained to what seemed to be a growing audience, is to somehow block the light from potential suns to make their planets and the atmospheres around them detectable. The fantastical occulter arrangement could make this possible. An intricately designed opening in the starshade would theoretically allow a space-based telescope some thirty-five thousand miles away to peek through and see only the tiny planets and, with an attached spectroscope, learn about its component parts. It would be Seager's longtime dream come true, and she spoke faster and faster as she described it.

The skies opened and drenching rain fell on the less-than-sturdy restaurant covering. Rain poured onto our table and onto those of other customers around us, forcing us all to crowd in together even closer where the awning still held. Seager left for a moment and a Dutch diner turned to me and, wide-eyed, asked for what I soon discovered to be a rather large collection of others, "Who is she?" I explained that she was an astrobiologist, part of an international effort to search for life beyond Earth. "I've just never heard someone talk like that," an American woman said. "Is all that stuff for real?" I assured them that it was.

Real enough that Seager has been on virtually every team put together by NASA or others to dig into and understand the science associated with that next step in planet hunting. She is on the science team of the Kepler mission (launched in 2009 with the goal of determining whether Earthlike planets are common in the more distant galaxy) and was on virtually every incarnation of NASA's Terrestrial Planet Finder effort. That mission, proposed and initially approved in 2004, would have launched two spacecraft designed to find signatures of life on exoplanets. Since 2006, however, it has been on indefinite hold, a victim of agency budget cuts, disputes within the exoplanet community on how to proceed, and extremely challenging and costly engineering and technology. In its place, NASA is considering a plan to design and launch an occulter and connect it with the James Webb Space Telescope, the highly sophisticated and very expensive successor to the Hubble Space Telescope and scheduled to launch in 2014. Seager is involved with that Webb-related effort, too.

That it might be possible to detect an Earth-sized exoplanet and then learn about the basic chemical makeup of that world illustrates just how far space science and technology have gone. In terms of scale, some have compared the feat of detecting an Earth-sized exoplanet to reading the mint date on a dime at a distance of more than three miles. What's more, the sun of an Earthlike exoplanet can be as much as 10 billion times brighter than the nearby viewing target. So the technological challenge is great and the costs high. Nonetheless, it's also true that the space agencies have made a hash of the effort.

In the early 2000s, as it was becoming clear that extrasolar planets were both plentiful and detectable, both NASA and ESA moved enthusiastically into mission-planning mode. At NASA, first one and then two high-end Terrestrial Planet Finder spacecraft were proposed and to some extent accepted, and ESA put money into a candidate of its own named Darwin. The logic of the missions was that the presence of certain elements and compounds is generally associated with biological activity, and that they could detect them on exoplanets using the most cutting-edge space technology. Oxygen, for instance, bonds quickly with many other elements, and so would be present in atmospheres as pure oxygen only if it were constantly being produced on the planet. The presence of ozone, a form of oxygen, would provide an even stronger sign of possible life. In certain circumstances, so would nitrous oxide and methane if they were present in substantial amounts; they, too, are frequently present when biological activity is going on. And, of course, the TPF missions would be looking for signs of liquid water. But NASA priorities changed a few years into the effort as human space exploration ate up larger amounts of the budget and disputes within the exoplanet community flared while cost estimates grew. By 2006, the exoplanet community learned that NASA had put the dream of taking that next big step regarding exoplanets not only on "hold," but on "indefinite hold."

As the big missions were losing support, two teams of exoplanet astronomers came forward with alternate, less ambitious and less costly plans. Both involved the seemingly far-fetched occulters, or sunshades, that Seager would later be describing so enthusiastically. Others had proposed similar techniques decades before, but exoplanets had not yet been discovered and the technology was very experimental, unwieldy, and complex, so it didn't go far. But things had changed by the mid-2000s, in part because a University of Colorado astrophysicist and space visionary, Webster Cash, had found a way to shrink the size of the sunshade from something miles long to something the area of a football field. As proposed, the starshade would be launched wrapped tight and then would be opened like a travel umbrella (using newly declassified technology) at its outer-space destination to a size of 170 or even 250 feet in diameter. The best shape to block the starlight turned out to be something akin to a sunflower or a daisy. The telescope would be 35,000 to 50,000 miles away from the starshade, but its aperture would be shaded by the distant barrier petal. And with the star's light blocked, the much fainter light coming off any exoplanets would be visible and the telescope could take the spectra of atmospheres and consequently find out what elements were present.

It was Cash who contacted NASA about the occulter idea in 2006, reaching out to Ed Weiler, who was then head of the agency's Goddard Space Flight Center, outside Washington. Weiler, who had previously been in charge of science for NASA and would resume that position a few years later, is known as an advocate for astrobiology and someone who believes strongly that life does exist beyond Earth. He's the same Ed Weiler who would later negotiate the NASA-ESA deal that will send methane-sniffing spacecraft and landers to Mars in 2016 and 2018. Weiler invited ten of his research scientists to the meeting with Cash, and he recalls being at first skeptical but then quite enthusiastic. That squares with Cash's clear memory of a moment in his PowerPoint presentation when Weiler realized he was proposing a plan that was much cheaper than the TPF, but capable of collecting much of the same information. "Ed jumped up and was pretty excited," is how Cash describes it. Soon after the meeting, Cash's Colorado team and another group at Princeton University won small but useful NASA grants to explore occulters. Northrup Grumman joined the Colorado team and actually built a small portion of an occulter as a limited proof of concept; Lockheed Martin joined with Princeton.

Both teams came up with what were considered to be feasible plans to use an occulter and a dedicated small space telescope together, and both were submitted in 2009 to NASA and to an arm of the National Academy of Sciences, which every ten years reviews all branches of space science and, with input from the scientists, sets priorities for the next decade.

Although both proposals were much less expensive than the original $6 billion TPF plan, they still were in the several-billion-dollar range and, because of NASA's budget problems, they were a long shot. Undeterred, Cash and others called a Hail Mary play—instead of sending up a telescope solely for the use of exoplanet searching with the sunshade, they would propose using the Webb Telescope, due to launch in 2014, as the second part of a resurrected occulter-telescope mission. The Webb is an enormous, $4.5 billion undertaking and has been in the works since 1996. Its main missions are to see further back than any telescope before to the time of the Big Bang, as well as to explore with greater precision the assembly of galaxies, the origin of stars and of solar systems. Viewing and understanding exoplanets is an important part of its mission, but it will not have the sophisticated equipment needed to analyze atmospheres or planet makeups. The idea of adding a new task at this late date seemed like a long shot, but Cash specializes in long shots. What's more, when he made his big advance in occulter technology in the mid-2000s, Cash first proposed joining forces with the Webb Telescope team and did some preliminary research before being advised to look elsewhere. He still thought the combination could work, and so did an important ally named Matt Mountain, the head of the NASA-funded Space Telescope Science Institute at Johns Hopkins University. The institute is NASA's organizing and data-analyzing center for the Hubble Space Telescope, and it will perform the same role for the Webb telescope, its successor. It's generally known as the hub of NASA's space-based observatory program, and is home to four hundred space scientists.

Mountain sought out a young, French-born research astronomer and specialist in starlight-blocking technology named Remi Soummer and asked if he was interested in looking into the feasibility of a Webb-sunshade mission. Soummer was eager to take up the challenge, even though his colleagues almost universally thought it was a waste of time for any number of technical, scientific, and bureaucratic reasons.

But a year later, when I met Soummer in his office in Baltimore, the wedding of the Webb to an occulter looked far more promising than anyone could have imagined. The scientific lead for the entire Webb project, Nobel laureate John Mather, supported the idea, as did the ESA principal investigator in charge of the instrument that would be used. (Both ESA and the Canadian Space Agency are partners on the Webb.) Soummer had just finished a white paper the week before (coauthored with Cash and Mountain, among others) outlining how and why the idea would work. There was, however, a big caveat: One minor filter needed upgrading on the Webb to fully shade the aperture and make the project work. It would be quite unusual for such a switch to be approved when a project was as far along as the Webb, but Soummer and others were hopeful. They had to be, because otherwise the characterizing of exoplanets, and consequently any deep knowledge of whether they had molecules and compounds associated with life in their atmospheres, would probably have to wait one or two decades, or more.

His explanation of occulters and the Webb gambit finished, Soummer pointed to a photo on his office wall. It was of a large dark globe with flickers of light around its edges. "That," he said, "is the ultimate occulter." The photo was of the moon passing in front of the sun in a total eclipse. It turned out that Soummer's interest in the field of occulters and other sun blockers, and his sense that the Webb team just might provide the needed filter upgrade, were tied up with events related to that photo.

He had taken the picture, and others that he eagerly brought out, back in 1999 when the total eclipse occurred in France. He was twenty-seven then and an astronomy graduate student in the southern French city of Nice. But he had done his homework and found that the area several hundred miles north of Paris would be the best place to view the eclipse. And so he and his brother headed up north and started to set up and calibrate their telescope at 3:30 A.M. for the noontime event.

Dawn came and it was cloudy. At 9 A.M. it was still cloudy, and the eclipse was looking more and more like a bust. But a while later Soummer saw a patch of blue in the sky some miles away and, with his brother, jumped into the car and raced toward the open sky. They went through small villages and past wheat and corn fields, speeding ahead in a way that Soummer now looks back on and cringes. When they finally reached the patch of blue, the eclipse was about to start. The two had about five minutes to frantically set up the telescope, to make sure it was pointing at exactly the right spot and was at the right angle. They were just finishing when the sun began to go behind the otherwise unseen moon.

"It was magic, the most beautiful thing I've ever seen," Soummer remembered. "It looked like dusk and then like night—a night with a full moon out—but it was in the middle of the day. I was looking and taking photos through the telescope and was always smiling." The eclipse, he said, lasted about fifteen minutes and he had a clear shot at it the entire time.

As he soon after learned, virtually nobody else in France had found their patch of blue sky, and so virtually nobody else had photos. He learned this when he took in his negatives to be printed, and a photographer friend told him he should try to sell them. The friend went to some of the big publications in France for Soummer, and ultimately found a buyer at *Science et Vie,* a French equivalent to *Scientific American.* They ran eight pages of Soummer's pictures, and he came away from the experience high as a kite. He also came away with a wonderful connection to the ultimate solar system occulter, and a sense that even very small odds of success can sometimes pan out.

Three months later, Soummer, Cash, and the Webb team were still deciding whether to switch filters to enable a future occulter mission. But the future for the occulter, or something like the occulter, looked bright. That's because the astronomy panel of the National Academy of Sciences, in its ten-year review of future priorities, placed exoplanets and their atmospheres at the very top of its influential list of scientific targets for the next decade. Writing that exoplanets are "one of the fastest-growing and most exciting fields in astrophysics," the report said that the goal of NASA and the National Science Foundation should be to "image rocky planets that lie in the habitable zone—at a distance from their central star where water can exist in liquid form—and to characterize their atmospheres." Many technical and bureaucratic and especially financial challenges remained, but the occulter plan now had a little blue sky. Searching for signs of biology on distant exoplanets might not have to wait a generation after all.

8 LIFE AND THE LAWS OF PHYSICS

We have learned a lot about how the universe came to be what it is now, but we know very little about why the 13.8-billion-year process played out in the entirely improbable way that it did. Astrobiology has happily moved into this vacuum, and not only asks the question "Are we alone?" but also the question "How are we here at all?" This is the kind of question that is generally considered outside the realm of science. But in this case, where the whole astrobiology effort is anchored in the expectation that a universe that led to our existence is likely to give rise to other forms of life, that non-scientific question of "why?" is impossible to entirely disentangle from the traditional scientific question of "how?"

To think through this question, astrobiologists, like scientists in other areas, have had to consider why the universe we know is so exquisitely fine-tuned. The concept refers to the well-established fact that life could never have started and evolved if the laws of physics were not almost precisely what they are. The more scientists learn about the cosmos, the greater the fine-tuning appears to be. Fine-tuning can be dismissed as a tautology (of course life arises only under conditions conducive to life), it can be embraced as an argument for a Creator, it can be seen as a series of signposts directing scientists to the deepest, least understood logic of the universe. But however it is interpreted, fine-tuning is a significant reality of the universe.

Here are some prominent examples:

- At the level of the cosmos, gravity is the organizing force. Yet gravity is extraordinarily weak compared with the electrical forces that hold together electrons, protons, and neutrons in an atom. That weakness, physicists have determined, is absolutely essential to the existence of our universe. A strong gravity universe would not only keep life from growing larger than a small insect, but would also pack stars closer together and with that proximity most likely keep stable solar systems from ever forming. Here's where the fine-tuning comes in: The ratio of the strength of the electrical forces in an atom compared with the force of gravity is 10 to the 26th. That's 1,000,000,000,000,000,000,000,000,000,000,000,000. As Lord Martin Rees, England's Astronomer Royal, described it, nothing as complex as humankind could have emerged if that number were even slightly smaller.
- The mass of a neutron in the center of an atom is 1.0013784 times heavier than the mass of a proton—in other words, they're virtually the same. This ratio allows atoms to remain stable and for chemistry to occur between elements. But if the proton were the same 0.1 percent heavier than the neutron, then the whole system would fall apart and life (and chemistry) as we know it would be impossible.
- British astronomer Fred Hoyle found in 1952 that the processes that form carbon, the indispensable element of Earthly life, depend on an improbable coincidence. His calculations, which have subsequently been confirmed many times, led him to conclude that very little carbon would be produced in the stellar furnaces unless the carbon nucleus vibrated at the same frequency as the nucleus of another element involved in the reaction, beryllium. Nobody even knew when Hoyle made his prediction that carbon nuclei could vibrate at that kind of frequency, but they soon confirmed that it did. And carbon, it turns out, is not only essential for life as we know it, but its presence in interstellar space is needed to cool down clouds of gases created by the dying explosions of large stars. Without that cooling, far fewer new stars would be formed.

The standard model of particle physics, which explains how atoms work, has about twenty constants that, if changed to any even minute degree, would make matter, stars, and galaxies very different and life, to a greater or lesser extent, impossible. Another ten cosmic constants order the universe. Some of these forces overlap, and not all require precise fine-tuning. But several do, and must be fine-tuned to an accuracy of greater than 99 percent to make a universe capable of forming and supporting life. It all sounds quite far-fetched, but it nonetheless is reality.

Fine-tuning has been hotly debated by physicists and cosmologists in recent decades, but without any real resolution. Since the descriptions of these physical relationships and interactions are demonstrably true, then the issue is not to prove or disprove them, but rather to make sense of them. Somehow our universe formed with physical laws, chemistry, and cosmic forces that allow for life on Earth and, quite possibly, elsewhere. How did that happen? Why did it happen? The Canadian philosopher John Leslie perhaps best conveyed the dilemma posed by fine-tuning with this parable: A man is facing a firing squad and fifty expert marksmen are preparing to end his life. The word is given and many shots are fired. To his amazement, the target opens his eyes after the fusillade and discovers he is still alive. What happened? Either he was stupendously lucky and everyone missed, or the marksmen intentionally missed their mark. As biological creatures in a finely tuned universe, we are that man.

In the search for other explanations, the theory of the "multiverse" emerged in the 1980s as a seriously studied alternative, although it was proposed as far back as 1895 by philosopher William James. Many variations on the theory exist, but all posit that we live in one of many, perhaps an infinite number of universes. Just as the earth is minuscule in comparison to our sun, so too would our universe be a speck in the enormous collection of universes that exist beyond our ability to detect them. Under the multiverse theory, countless universes exist where the necessary forces did not combine in a way to allow for life, while leaving room for the possible formation of a universe like ours where they did. The theoretical logic is strong, but some scientists argue the multiverse idea is not actual science since it can't be either verified or falsified now, and perhaps never will be. Since the multiverse presupposes universes in many different dimensions, at distances farther than the speed of light could travel in the 13.8 billion years of our universe, the ability to tease out the reflected presence of a second or third or billionth universe is absent. Until some way is found to detect another universe, the multiverse can only remain a plausible if unproven theory or, even worse, speculation—even though the number of physicists and cosmologists who embrace some variation of the theory grows ever larger.

Multiverse thinking—the attempt to address fine-tuning and other questions of theoretical and cosmological physics—proposes many different kinds of universes and dimensions. There's the bubble universe theory, which assumes that our universe as well as numerous other universes were formed from a "bubble" of a "parent universe." The many-worlds interpretation posits only one universe, but it splits into "many worlds" based on the logic of quantum mechanics. These worlds, however, cannot interact with each other. The so-called strong anthropic principle, in one of its interpretations, says that a range of different universes is necessary for the existence of our own. It also says, however, that the universe exists because we are here to observe its existence. None of these approaches has attracted anything close to a scientific consensus.

Theoretical physicist Lee Smolin, a founding professor of the Perimeter Institute for Theoretical Physics in Waterloo, Ontario, has sought to reconcile the concept of fine-tuning with science by using the idea of a cosmic natural selection. His best-known work, *The Life of the Cosmos,* makes the case that many of the seemingly fine-tuned aspects of the universe can be explained by a kind of cosmic Darwinism—one in which differing laws of physics in effect compete and change over time, allowing them to evolve in a way that leads to a finely tuned world. (I liked Smolin even better when I saw a video of a lecture he gave at his institute on the physics of the universe. The pointer he used was a clunky wooden hockey stick.)

Since the mid-1980s, when he studied biology as well as theoretical physics, Smolin's pathway into understanding and working to resolve the fine-tuning problem has focused on the parallels he intuited between the two fields. Those perceived parallels led to his theory that the universe, like the Darwinian world of life on Earth, evolves under the pressure of natural selection. The rise of life and ultimately humans is not the goal of that cosmic process—any more than humans are the goal of Darwinian evolution—but it is a predictable offshoot. As Smolin explains it, the key is star creation. Not a new idea, but he adds a twist: that the very same processes that lead to the existence of long-lived stars happen to support the almost infinitely lengthy trail of processes essential to make biology possible. That mutuality of results is, he says, "an important clue for fundamental physics." So too is the abundance of carbon dioxide and oxygen in the universe—not because it has anything to do with life per se, but because it helps accelerate or increase the formation of massive stars that then give rise to much greater molecular complexity, which in turn makes life possible far down the road. "So the universe," he argues, "evolves in ways hospitable to life as part of natural selection, the movement towards a more complex universe." Fine-tuning, in this interpretation, is the process by which more powerful, more fit laws of physics triumph over others, and life follows in the wake. Smolin said that he has tried to apply his cosmic natural selection theory to a single-universe model, but so far "couldn't find an approach that didn't yield predictions that disagree with experiment."

This is a very short version of a long scientific story that includes a universe where the laws of physics can be different in disparate regions, where the existence of other universes is mathematically essential, and where black holes may be the key to the formation of those other universes. The idea that black holes, where the laws of physics fall apart under the pressure of concentrated gravity, might play a central role in the formation of new universes cannot be proven and has many detractors. But Smolin argues that black holes are nurseries for Big Bangs, that the collapsing in of matter into black holes leads directly to the formation of new universes on the other side. Each universe will have a different set of laws of physics that either can or cannot evolve into a structure that supports life.

Cosmic natural selection is a long way from being proven, but Smolin says it can indeed some day be proven or disproven, unlike other multiverse theories. Also, it offers an explanation of how fine-tuned physical laws and ultimately life could arise in a universe that isn't "designed" to do either. It just happens, in a way similar to how a single-cell bacterium over the eons evolves into an elephant. The theory brings life into a scientific cosmology, and it extends astrobiology far into the cosmos. We also have a well established parallel to learn from—Darwinian evolution.

"A long time ago, your ancestors were fish," writes Paul Davies, another iconoclastic physicist and cosmologist, British-born but now director of the Beyond Center for Fundamental Concepts at Arizona State University, where he writes prolifically about the logic of the universe and big-picture astrobiology. "Think how fish spawn countless eggs, and imagine the tiny, tiny fraction that survive and mature. Nevertheless, not one of your ancestors—not a single one—was a failed fish. What are the odds against this sequence of lucky accidents extending unbroken for billions of years, generation after generation? No human lottery would dare to offer such adverse odds. But here you are—a winner in the great Darwinian game of chance! Does this mean that there is something miraculous in the history of your ancestry? Not at all."

Can't the same be said of the ancestry of the Earth and its menagerie of life? Or of life elsewhere in the universe, or universes?

There is, however, an alternative view, one that involves a Creator, and it is held by some sophisticated scientists. A long tradition exists of attempting to support the existence of a Creator through science (Newton, Copernicus, and, more recently, Cambridge physicist-turned-Anglican cleric John Polkinghorne come to mind), but the track record has not been good. Nonetheless, Smolin, who certainly does not subscribe to the Creator view of the origin of the cosmos, suggested that I speak with South African theoretical physicist George Ellis, who does hold that position.

Ellis, born in 1939 and now retired from the University of Cape Town, has credibility because he coauthored a seminal book on cosmology with Stephen Hawking (*The Large Scale Structure of Space-Time*) and because he returned to South Africa after his studies at Cambridge University and joined the nonviolent wing of the fight against apartheid. Neither means he's correct about the nature and meaning of fine-tuning, but he is nonetheless well regarded in the field. A practicing Quaker, he has published prolifically on cosmology, multiverses, and fine-tuning, and generally argues the following: The fine-tuning of the universe makes it likely that life in our universe is common, that no scientific proof has been offered (or probably ever can be offered) of the existence of universes beyond ours, and that the existence of multiverses is as much a question of belief as the existence of a Creator.

"The more we learn of the universe, the more we learn how great the fine-tuning really is," he told me. "Since science cannot tell me that any of the various explanations for that reality is true or false, then a plausible hypothesis is that of a Creator. It's not provable, but nothing else is, either."

His long-ago collaborator, Stephen Hawking, sees the same fine-tuning and comes to a very different conclusion, one that did not sit well with some of the more religiously minded. In his 2010 book, *The Grand Design,* written with California Institute of Technology physicist Leonard Mlodinow, Hawking describes the universe (or universes) as ultimately understandable by science—a view shared by quite a few in the field.

"Our universe seems to be one of many, each with different laws. That multiverse idea is not a notion invented to account for the miracle of fine tuning. It is a consequence predicted by many theories in modern cosmology," he writes. "As recent advances in cosmology suggest, the laws of gravity and quantum theory allow universes to appear spontaneously from nothing. Spontaneous creation is the reason there is something rather than nothing, why the universe exists, why we exist. It is not necessary to invoke God to light the blue touch paper and set the universe going."

A strong statement for sure, but notice how many qualifiers are written into even Hawking's explanation. The issue is far from settled and few scientific challenges are as great as those posed by the fine-tuning of the universe. But few hold greater potential for explaining how and perhaps why we are here, and why other life-forms in the universe might be out there, too.

Just as astrobiology is inevitably drawn into the worlds of cosmological fine-tuning and multiverses, so too is it being pulled into an equally fantastical world here on Earth—that of a possible shadow biosphere that supports life with a different origin and different characteristics than our own. Science has never found any alternate life-forms, proponents say, not because they don't exist, but because scientists have never looked for them.

That has begun to change. Nothing is for certain in their work, but a handful of researchers have made some intriguing discoveries that suggest a shadow biosphere just might be present. What began as a theory is now the subject of NASA-funded work at hypersalty, hyperalkaline Mono Lake in California, about one hundred miles north of Death Valley. A terminal lake that receives water from the nearby mountains, it has no outlets and so only loses water through evaporation (until the city of Los Angeles began siphoning off water in 1941). Mono Lake is known for the "tufa" columns of limestone that stand in its midst and give it a distinctly spooky quality, as well as a very unusual chemistry caused by its lack of outlet streams. That means elements and compounds that pass through other lakes and get dispersed into big rivers and later oceans stay put in Mono Lake and concentrate to abnormally high levels. Arsenic from the nearby Sierra Nevada flows into Mono Lake and stays—creating a toxic stew with arsenic levels seven hundred times higher than what the Environmental Protection Agency considers safe. Despite being a virulent poison for most living things, arsenic has emerged as the key element in shadow biosphere research. In fact, if the research holds up to the critiques it has attracted, it will represent the beginning of a new era of biology—one where the already fuzzy concept of life as we know it will get much fuzzier.

The main force behind the arsenic biosphere research is a thirty-three-year-old biochemistry whiz named Felisa Wolfe-Simon. She broke onto the NASA scientific scene in 2008 when she attended an exclusive Gordon Conference meeting on "The Origins of Life" and raised the possibility of life-forms on Earth with chemical makeups that are entirely incompatible with all other life that we know. At the time, she recalls with something between pride and dismay, her mane of dark hair was dyed bright pink, and she sported a number of piercings. That probably didn't help her establish early credibility.

But over the next four years, she attracted the attention of a number of top scientists, ranging from biologists to cosmologists. She worked with geochemist Ariel Anbar at Arizona State University and he introduced her to Paul Davies, the unconventional physicist/astrobiologist (and prolific writer), who already had a strong interest in the "shadow biosphere."

Davies has promoted the shadow biosphere idea for some time, as had University of Colorado philosopher and astrobiologist Carol Cleland, who actually coined the word. His argument is part scientific, part practical. Why spend billions on flying to distant planets in the hope of finding evidence of current or former life different from ours when it may well exist right under (or in) our own noses? Many origin-of-life scientists assume that life didn't begin just once on Earth, but rather a number of times in similar but nonetheless distinct forms. The organisms that weren't based on carbon, nitrogen, and phosphorus perhaps couldn't compete as well and died out, or maybe remnant populations live undiscovered because nobody has ever looked for them. But now they're looking.

Davies and Wolfe-Simon submitted a proposal on an arsenic-based shadow biosphere to the John Templeton Foundation in 2007, but the request was turned down. A prime reason why was that one of the reviewers, a senior arsenic specialist at the U.S. Geological Survey in California named Ron Oremland, didn't believe there was sufficient reason to think the research would be successful. But Oremland was nonetheless intrigued. He had spent much of his career, after all, studying the interactions between arsenic compounds and surrounding biology, and he felt a little guilty that he had panned the proposal. He ran into Wolfe-Simon several more times in the next few years and then opened his USGS lab to her so she could focus on Mono Lake as the location for a possible shadow biosphere.

But that required outside funding, which ultimately came in the form of a fellowship from NASA's Astrobiology Institute. She headed to California, started collecting mud from the lake, and began the tedious process of sifting and concentrating samples containing already high levels of arsenic. She then began to examine the microorganisms that made their living in the toxic environment, and found something unusual. All known living things on Earth contain the elements carbon, hydrogen, nitrogen, sulfur, oxygen, and phosphorus, which forms the backbone of all genetic material and, in the form of the molecule adenosine triphosphate, is essential for energy storage and transfers in cells. Yet it appeared that some of the microbes from Mono Lake could survive with little or no phosphorus in them, while having very high levels of arsenic.

Arsenic is chemically very similar to phosphorus, a downstairs neighbor in the table of elements, and its toxicity is in large part a function of the fact that other molecules initially mistake it for phosphorus and then are destroyed when the difference is revealed. But the microscopic Mono Lake organisms—from the domains of bacteria and archaea—not only withstood the arsenic but seemed to be possibly using it as a substitute for phosphorus, which, along with carbon, oxygen, hydrogen, and nitrogen, are the key and essential elements of life on Earth. Through months of lab work, Wolfe-Simon and Oremland grew Mono Lake samples with higher and higher levels of arsenic until they reached a point where arsenic had replaced a significant percentage of the phosphorus and arsenic levels were some forty thousand times the EPA safe level. Yet some microbes survived when fed glucose and vitamins, as evidenced by how the water slowly became cloudy with biological activity.

The samples were then sent to several of the nation's best labs with the most sophisticated equipment for molecular-level testing, and the results were startling: The arsenic, they found, was incorporated into the genetic material (the DNA and RNA) of the cells as well as essential proteins and the cell membranes.

Word of the potentially ground-breaking discovery was first announced in an embargoed release from the journal *Science*, which was followed by the NASA public announcement of an upcoming press conference to discuss a finding that could have implications for extraterrestrial life. The news shot through the blogosphere, with detailed predictions of life on Jupiter's moon Titan, or a new day in extraterrestrial research. Both *Science* and NASA remained silent for four days until the press conference, which by then was anticipated to be news on the scale of the 1995 Mars meteorite announcement or greater.

The press conference focused on the findings in the *Science* paper—that microbes from Mono Lake could be grown with lots of arsenic but virtually no phosphorus, and that sophisticated technology had been used to find that the arsenic was contained within the DNA and other essential genetic and life-supporting molecules of the microbe. While the result was remarkable, the larger take-home message was even more so. "We have cracked open the door to what is possible for life elsewhere in the universe," Wolfe-Simon said. Ed Weiler, NASA's associate administrator for the Science Mission Directorate, was not at the press conference but did add this in a formal release: "The definition of life has just expanded. . . . As we pursue our efforts to seek signs of life in the solar system, we have to think more broadly, more diversely, and consider life as we do not know it."

The results were presented with the proper caveats—that they had to be confirmed and expanded upon—and respected chemist and astrobiologist Steve Benner was also onstage to make a strong case for the near impossibility of the substitution of arsenic for phosphorus in DNA. Arsenic breaks down quickly in water, he said, while phosphorus does not. So how could arsenic bonds hold up in an aqueous environment?

To say the scientific blogosphere was skeptical would be an extreme understatement. Some bloggers immediately attacked the research as incomplete or incompetent, and others concluded the results were simply impossible. Many dinged NASA for "hyping" the discovery, and the peer reviewers at *Science* were dismissed as compromised. Some of the critiques and challenges were sincere and based on science, but many were personal and nasty. The Mono Lake researchers had predicted a heated response, but they were taken aback by the venom. Perhaps they shouldn't have been. History tells us that developments related to astrobiology and the search for extraterrestrial life bring out intense emotions. And NASA did put out a release the day of the press conference, saying that agency-funded "astrobiology research has changed the fundamental knowledge about what comprises all known life on Earth." What was an historic and proud moment for NASA and the researchers was a red flag to many others.

For the first week, editors at *Science* and the researchers were silent except to say they would address challenges and critiques through the traditional peer review process. But after two weeks, the blogosphere was sufficiently livid that all felt the need to respond—to address some specific charges and to make clear that the microbes would be made available to anyone who wanted to test them in their own labs. In other words, they acknowledged the criticism but held their ground. Their research had been peer reviewed on several levels and so had already been challenged and challenged again. Nonetheless, Wolfe-Simon's colleague at the U.S. Geological Survey, Ronald Oremland, joined a panel set up at an American Geophysical Union annual meeting specifically to discuss the controversy, as opposed to the science. An old-school scientist, who said he was trained to discuss his work in journals and at conferences rather than on the web, said he had not responded online because, "You can wind up in a Jerry Springer situation before you know it, with people throwing chairs." But if the controversial research holds up and further investigation supports the finding that the arsenic really is woven into the microbes' genetic material, then we'll be in uncharted waters. Microbes with arsenic instead of phosphorus in their DNA backbones would not just represent another interesting discovery. The work of Wolfe-Simon and her team would open a new window into life—an alien life, if you will, right here on Earth.

9 FAR-FLUNG INTELLIGENT WORLDS

The countdown was set to begin at Shin-ya Narusawa's mission control room at the Nishi-Harima Astronomical Observatory in southern Japan. Nobody was going out into space; no spacecraft were being sent into orbit. But it was a noteworthy night because another nation was joining the improbable yet increasingly sophisticated search for extraterrestrial intelligence elsewhere in the galaxy.

Jumping up to suddenly rigid attention, Narusawa called out "T minus ten." The assembled reporters and camera crews jumped up too and surged toward him.

Narusawa, the chief researcher of the observatory in the hills northwest of Osaka, had eyes red with fatigue, but his smile radiated triumph. For more than two years he had been preparing and organizing for this moment and now it was about to begin: Japan's first coordinated, all-country observation of a single star system that might, just might, be home to intelligent beings sending signals out into space. Narusawa said he wasn't 100 percent sure of this, but it might be the first large-scale, simultaneous SETI observation of its kind in the world.

"T minus eight minutes," he called out. The star system that the thirty radio and optical telescopes were focused on was one recommended two decades earlier by none other than Carl Sagan and Harvard University's Paul Horowitz, then and now a leading proponent of the search for extraterrestrial intelligence, or SETI. As a young man, Narusawa was enamored and inspired by Sagan and his work—he read the book *Cosmos* at least five times—and now he was about to launch his own major SETI observation based on guidance from the masters. Their "Megachannel Extra-Terrestrial Assay," or META, spectrum analyzer had a capacity of 8.4 million channels and the ability to use Doppler shift to distinguish between terrestrial and possible extraterrestrial samples. The project, begun in 1985, was led by Horowitz with the help of the nonprofit Planetary Society and with financial support from movie director Steven Spielberg.

Getting to countdown had not been easy. The day before, Narusawa had explained with evident emotion that the observation would be named "Sazanka," after a light purple shrub flower that comes out in the fall, the rainy season around the observatory in Sayo. He chose Sazanka because, in Japan, the plant represents the ideal of "never giving up"—a designation earned by the way that the flower survives the cold, even after snowfalls. The choice also carried a small, inside joke: The rainy season is the high time for *sazanka,* when it flourishes, but is a low time for the observatory because the clouds make it much more difficult to get a good night's observation. Several banners with images of the *sazanka* were hung around the control room, and one perfect flower was placed in a small glass vase near Narusawa's chair. All around him was the message: "Never give up." Given both the difficulty he had experienced devising and planning the effort, to say nothing of the needle-in-a-haystack chance that he would succeed where no other SETI observation had for fifty years, the *sazanka* seemed an apt symbol of encouragement.

"T minus six minutes," he proclaimed. Narusawa had begun his optical SETI observations in 2005, after reading that Horowitz himself had begun a program using an optical telescope for SETI purposes, rather than the traditional radio telescopes used since the beginning of the SETI effort in 1960. The Nishi-Harima telescope that Narusawa controls is the largest optical telescope in Japan, and he was looking for an exciting and unique additional use for it after it came on line six years earlier. Optical SETI looks for nanosecond flares or lasers coming from a distant star system, just as the far more widespread radio SETI listens for radio messages or bursts that intelligent creatures might be sending. Optical SETI had long been seen in the field as useless because neither it nor laser technology was sufficiently refined. But now it is.

Narusawa, briefly seated, jumped up again and declared: "T minus four." Playing in his mind at that moment was one of those long-ago, formative moments of his life. He was watching live on television an early Apollo moon walk; or, rather, a series of goofy Apollo moon jumps since the astronaut's movements looked to him exactly like a kangaroo hopping. He was beyond delighted—he was inspired and decided then and there to study astronomy and learn about the universe. If his memory is correct, the forty-four-year-old Narusawa was then four or five.

"T minus two," he called. Nishi-Harima is on a mountaintop, or perhaps more accurately a hilltop, outside the town of Sayo and quite far from the major cities of Japan. So the fact that six television stations and several newspapers had sent reporters out to record the event showed a not insignificant interest. As in the United States, curiosity (and skepticism) about advanced extraterrestrial life is always present, if not terribly sophisticated. It's more the stuff of science fiction, comics, and horror films than of real science in the public mind. But Narusawa is one of many trying to change that. He believes that searching for intelligent extraterrestrial life is science at its best, science "serving humanity" in the way it should but often does not. In addition to the high-mindedness, he also knows searching for ET can be popular, can engage the public if done with the right pizzazz. Narusawa has some of the showman in him, and he's not shy about it. He would like to be Japan's Carl Sagan.

"T minus one." Over the whole length of Japan, some thirty observatories, university telescopes, and amateur sky watchers were set to turn on their machines, having set them to the precise coordinates Narusawa had ordered: a star system in the constellation Cassiopeia, just where Horowitz and Sagan had proposed years before, and where a Japanese radio astronomy colleague, Mitsumi Fujishita of Tokai University, had detected an unusual and unexplained blip in his monitoring four years before.

"Start observations," Narusawa finally declared. He was speaking, of course, to the assembled journalists (and for that last order he was actually speaking in Japanese). The other astronomers couldn't see or hear him, but on their own knew the big moment had arrived. Thirty monitoring points began to simultaneously look at and listen to the exact same spot in space, hoping to detect some signal—a burst of light, a radio signal—that could only be made by intelligent life. They would all continue for two full nights, weather and clouds permitting.

Scanning for extraterrestrial communication is for the very patient and the very determined—the *sazanka* of astronomers. Narusawa would not have any results for months, and ultimately they would not find any sign of ET life. Millions of similar hours have been spent by SETI scientists and advocates around the world—most of them in the United States—and so far no radio blip or laser burst has been confirmed. Nonetheless, SETI observations have been spreading around the world. South Korea, Italy, Argentina, and Australia also have operating SETI programs, and American institutions including Harvard and Princeton have geared up substantially as well, along with the now-flourishing SETI Institute in California.

There's a simple reason for this, and it has nothing to do with any perceived breakthrough on the horizon, although the technology to hear or see communications from distant planets is in fact advancing and expanding rapidly. It's rather a function of the near certainty that no intelligent life exists in our solar system except on Earth. There may well be microbial, bacterial life to be found on Mars, Europa, Enceladus or Titan, a second genesis that would strongly suggest life, even complex life, in the universe is a commonplace. But if the explorations of the modern space age have shown anything for sure, it's that conditions for the evolution of what we understand to be intelligent life do not exist elsewhere in the solar system. The planets and moons are either too hot, too cold, too dry, or made up primarily of gas.

So if scientists are ever going to make contact with the intelligent life many of them are convinced is out there, it will have to happen via interstellar communication, since earthlings are not going to reach the warp speeds necessary to reach other galaxies anytime soon. The nearest star to our own is Alpha Centauri at 4.3 light-years away, and the nearest one with an identified terrestrial exoplanet is Gliese 581, at about 20 light-years away. That's about 704,952,200,808,876 miles.

This may seem like an enormous, almost certainly impossible wall to climb. But for astronomers familiar with the vastness of the universe, the minuscule size of our own world, solar system and even Milky Way galaxy, the improbability lies elsewhere. To them, it defies scientific logic to think there isn't other intelligent life out there. So for Shin-ya Narusawa and scores of others around the world, the vanishingly small chance of detecting and confirming a signal from a distant civilization is science at its most exciting—it's a wide-open field, it's challenging on many technical levels, it remains unconventional and even controversial, and it holds the promise, however faint, of someday making what would be the biggest scientific discovery of all time. Substantially larger telescopes are being planned and constructed and the logic of Moore's law, which says the speed of top-notch computers will double every eighteen months or so, offers hope for a much faster and better read of the mountains of data produced by the science instruments attached to those telescopes.

SETI, or in this case OSETI for its optical version, makes additional sense in Japan because of the way telescopes are perceived and paid for. Narusawa's Nishi-Harima observatory is the center of a large, hilly park that includes ball fields, hiking trails, and guesthouses, and was funded largely by the Hyogo prefecture. Astronomy is certainly science, but it's also part of "culture and recreation" as far as Japanese policy makers are concerned. The instruments, including the $40 million Nayuta telescope, always reserve time for local people to come in and look at the nighttime sky. "In Japan, our telescopes are all open to the regular people, and when they come in we want to know what are their big interests in astronomy. The top two are these: Is there an end, a border, to the universe? And is there life, especially intelligent life, anywhere other than Earth? So OSETI is what people want." Narusawa said his goal, his dream, was conducting observations with SETI programs in the United States. And not quite a year later, after Narusawa had presented his Sazanka data at a large, annual astrobiology meeting in Texas, SETI officials agreed to joint American-Japanese observations. Those soon grew to include observatories from thirteen nations on five continents, and a "Project Dorothy" global SETI took place in late 2010, with Nishi-Harima as headquarters. Narusawa was beside himself.

The United States is where SETI was born and where it now—despite decades of skepticism—flourishes. The Japanese team showed impressive coordination, technology, and determination, but they were pretty much groping in the dark. That's a higher-tech but nonetheless similar place to where SETI pioneer Frank Drake was in 1960 when he first aimed the newly inaugurated antenna of the National Radio Astronomy Observatory in Green Bank, West Virginia, at two relatively nearby star systems, and set about listening for messages. His project Ozma (named after Princess Ozma of Oz) lasted but two months and ended without any contact. But that hardly mattered; a dream was born.

Almost a half century after Drake began his work, the conceptual and technological offspring of Ozma in America are busy listening for that same fleeting signal. Laid out across a high valley between California's 14,000-foot Mount Shasta and 10,000-foot Mount Lassen, forty-two radio dishes with enclosed antennas stand sentinel and collect data in a systematic way hardly imaginable when SETI began. (This is quite literally true: Drake says the array now works at a level of effectiveness something like 100 trillion times greater than what he had in Green Bank.) A late fall snowstorm was moving into the Hat Creek area when I arrived, and the radio dishes were groaning and grinding in the wind—a perfect Earthly backdrop for an otherworldly venture. But until the winds reach a sustained 30 miles per hour they continue their job—a sometimes targeted, sometimes full-sky, systematic, and nonstop effort to hear radio communications from afar. The telescopes in operation, the biggest radio telescope array in the world, increase by a factor of at least 100 the ability of SETI searchers to find the kind of "transient" radio signals they're looking for in the sky. But it's what SETI hopes is only a beginning. Plans for an enlarged Allen Telescope Array at the Hat Creek Radio Observatory call for 350 radio antennas and ever-faster computing power, a concentration of dishes and Moore's law upgrades that will increase the ability to detect and make sense of radio signals by a factor of 1,000.

The array is set in a leveled field surrounded by lava beds, red fir, and snowcapped mountains, and redolent with the sweet smell of sage even as the snow was falling. Most of the near neighbors walk on four legs. The area, and especially Shasta, looms large in many New Age spiritual tales, and has been considered a sacred spot to native inhabitants going back many centuries. The area has also long been known for reports of Sasquatch or Bigfoot sightings—all part of a worldview the SETI folks are careful to keep far removed from their endeavor. Most of the actual data monitoring and analysis is done far from the field, where astronomers (and their graduate students) receive the data coming into Hat Creek via computer and do their work.

It was only when the strong gusts calmed a bit that I could tell that the wind had camouflaged the sound of the programmed, periodic movements of the twenty-foot-diameter dishes. With a choreography both graceful and surreal—elephants dancing ballet—they moved in pairs or groups to aim at a different star or whole other galaxy. This morning's mission was to focus on 100 of the 500-plus extrasolar planets discovered in the past fifteen years. It would be the first SETI endeavor of its kind, but followed the logic that animates astronomy (and much of science). A discovery rearranges the scientific furniture, and then researchers in related fields start using the new reality for their own purposes.

The wind was picking up and gusting hard again as I came across one of the numerous antennas named for accomplished donors. This one was named jointly after Jack Welch—not the former chairman of General Electric and business seer, but the famous radio astronomer—and his wife, Jill Tarter. It was a serendipitous but appropriate introduction to the story of how this unlikely, faraway field in the Lassen National Forest came to host 42 radio antennas and may someday have 300 more—all of which would be listening for messages and signals that alien civilizations just might be sending our way.

Jill Tarter, as all SETI enthusiasts know, is a founder, and now the matriarch, of the SETI Institute. A passionate visionary of extraterrestrial intelligence, she became a rather high-profile cultural figure after release of the 1997 movie *Contact,* which was based in part on her life. (Jodie Foster played her in the movie, which regrettably did justice neither to the subject nor to the individual, although Tarter says it worked miracles in terms of fund-raising.) An engineer and astrophysicist by training, Tarter has an endowed chair at the SETI Institute, located in the high-end Silicon Valley town of Mountain View, and just across Route 101 from NASA's Ames Research Center, which once was involved in SETI work as well. Jack Welch may not be quite as well-known in popular culture, but in the world of astronomy he is a giant, too: the first to find water vapor and organic formaldehyde in distant space, discoveries that rewrote the textbooks about the makeup of interstellar space. He led the radio astronomy lab at the University of California, Berkeley, for twenty-four years, and since 1998 has had an endowed chair as well, the Watson and Marilyn Alberts Chair in the Search for Extraterrestrial Intelligence, in the astronomy department at Berkeley. It was the first such chair, and remains the only one of its kind. Given their highly specialized skills and unusual interests, Tarter and Welch were destined to meet. They not only met but they both eventually divorced and re-married.

Each is a forceful personality on his or her own; together they are a power duo that can make things happen. In the late 1990s, SETI and Berkeley formed a partnership to build the array, and in 2004 construction began at Hat Creek, thanks to an initial $11.5 million donation from Microsoft cofounder and billionaire Paul G. Allen. Pleased by what he saw, Allen later put in another $13.6 million, and the initial forty-two-dish array was dedicated and began work in 2007. While the telescopes would fulfill the decades-old SETI dream of having cutting-edge equipment that was dedicated to its goals, it would simultaneously provide equally cutting-edge technology for the Berkeley radio telescope program, which researches more traditional and incrementally revealed subjects such as how galaxies form, the nature and properties of dark matter (ubiquitous in the universe but known only by its gravitational pull), and the nature and workings of black holes. Under Welch's leadership, the array is also drawing a cosmological map of the presence of hydrogen in the universe.

Hat Creek represents quite a coming of age for SETI, which long struggled for telescope time, funds, and respectability from the federal government, and ultimately secured none in substantial or dependable form. The giggle factor was just too high. During the 1970s and '80s SETI was funded to a limited extent by NASA, and in the early 1990s was finally embraced and awarded funds for a more sophisticated program. But that dream ended quickly in 1993 after Senator Richard Bryan, Democrat of Nevada, made zeroing out SETI into a personal priority. "The Great Martian Chase may finally come to an end," he said as he closed in on his goal. "As of today millions have been spent and we have yet to bag a single little green fellow. Not a single Martian has said take me to your leader, and not a single flying saucer has applied for FAA approval."

The odds remain long that Hat Creek and other SETI efforts will make contact with distant civilizations, but what the scientists are doing is far from the caricature presented by the senator from Nevada. The former senator would be dismayed to know that not only has SETI attracted millions from savvy high-tech entrepreneurs and scientists, but the Institute is again eligible to compete for NASA funds after its program was deemed to be scientifically sophisticated and sound. The forty-two dishes are unusual not only because of their SETI mission, but because they're also a cutting-edge design and generally seen as the future of radio astronomy.

The United States already has one huge radio astronomy dish at Arecibo, Puerto Rico, which is a thousand feet across and can pick up radio signals from 100 million light-years away. For years it was the primary site for SETI work, though listening time was limited. A much-used "giant eye" with real star power—it has served, after all, as a backdrop to movies ranging from *Contact* to the James Bond film *GoldenEye* and an *X-Files* episode called "Little Green Men"—it nonetheless belongs to the past and is struggling to hold on to the limited National Science Foundation funding that it receives. Not only would it be prohibitively expensive now to build an Arecibo, but it would be seriously behind the technological curve. Welch in particular pioneered the notion that many smaller dishes hooked up together would have the same power and sensitivity as a big dish, and would do it at a much lower cost. Hat Creek is the prime example of that change in approach and vision, and its receiving dish is actually as large as the distance between the most distant antennas. At 350 antennas, the Hat Creek dish would be, in effect, a kilometer in diameter.

Space technology aside, Hat Creek also gives the SETI enterprise a permanent and easily expandable home. Over the next two dozen years, the Allen Telescope Array as currently configured will gather a thousand times more information from distant star systems than has been collected in the past forty-five years. The data will also be far more precise and will come from fainter, more distant stars. Perhaps most important, the dishes produce a detailed image of a broad expanse of the radio sky at any given moment. This is how Seth Shostak, SETI's lead astronomer, put it: "Let's say you're looking for elephants in Africa. If you have one guy with binoculars, your chances of seeing a tusker are pretty limited and it will take a long time to succeed. Compare that with having one thousand guys with binoculars looking for elephants. Suddenly, you'll be finding a lot of elephants."

The Hat Creek radio waves are collected, bundled, and carried to waiting computers that split them into four categories based on their radio frequency. Generally, two of the bundles go to SETI and two to Berkeley, a sharing that made the Allen Array so appealing in the first place. SETI needed the solidity of the Berkeley lab to be credible when it approached potential donors, and Berkeley needed the sexiness of SETI. With both Jill Tarter and Jack Welch convinced it should go up at Hat Creek, the momentum was hard to stop.

The antennas are designed and programmed to pick up distant radio waves (a relatively low-energy form of radiation) at a frequency between 1,420 megahertz and 1,660 megahertz. That's a region of the radio end of the electromagnetic spectrum that has less competing background "noise" from the universe than most others and, as a result, more radio waves, and more distant radio waves, can be detected. In radio astronomy and SETI parlance, that range is called the "Cosmic Water Hole," because it exists between the points on the spectrum where radio waves from interstellar hydrogen (H) and interstellar hydroxyl ions (OH) arrive on Earth. It was Bernard "Barney" Oliver, a top engineering executive at the Hewlett-Packard Company, the early head of the NASA SETI program and later the first president of the SETI Institute, who gave the "water hole" its name. The logic was simple: The spectral region between H and OH was relatively tranquil and quiet, rather like water (H_2O), and the relative calm seemed obviously to be a function of the breaking apart of interstellar H_2O. Oliver proposed that any advanced civilization would be able to similarly identify the "water hole" as a good frequency for communications, and the scientific-extraterrestrial community agreed. "Where shall we meet our neighbors?" Oliver famously asked decades ago. "At the water-hole, where species have always gathered." Many SETI observations are still done at Hat Creek and elsewhere at water-hole frequencies.

As Hat Creek station scientist Rick Forster explained it, a major reason for choosing Hat Creek as home to the array was that it was especially quiet in the "water hole." Many other kinds of man-made radio noise now pollute that region of the radio spectrum, and it's increasingly hard to find places where that new noise doesn't drown out signals that might be coming from far away. Locations like Hat Creek, the scrub deserts of New Mexico and Arizona, and the high Atacama in Chile are home to the best new radio arrays because—while they still have to account for radio waves coming from cell phones, high-definition television, and military satellites—there is far less radio noise from surrounding human activity than most places.

The SETI Institute is always struggling for money, but the overall enterprise is more capable and stable now than ever before. So, fifty years into SETI, it seems entirely fair to ask this question: Why, if scientists and ET enthusiasts are correct and there are many technologically advanced civilizations in the universe, have we neither heard nor seen any signs of them? Earth would plausibly be the kind of place intelligent extraterrestrials might try to contact, since its atmosphere has been awash in probably the most telling signatures of life—oxygen, ozone, and water—for 2 billion years. For the last hundred years, we've also put out an enormous amount of radio traffic that could plausibly be detected from deep space. Yet there have been no incoming messages detected. Some in astrobiology see SETI as the field's most vulnerable Achilles heel. Too much looking and listening; not enough finding. As a result, the half-century-old SETI policy of relying primarily on radio telescopes to listen for messages is frequently challenged. So, too, is the SETI decision to remain in a listening mode rather than sending out pings and flares from Earth that might inspire a response. SETI scientists say listening offers the greatest chance for success, but increasingly, others disagree.

But ask Frank Drake whether the absence of contact means we're alone in the universe as the sole intelligent creatures, and he laughs. The assumption that underlies the question, that we've done enough SETI monitoring that something definitive should be known by now, is so far off base that to Drake it's much more humorous than it is threatening. The author of the 1963 "Drake Equation," which sought to very roughly estimate the number of advanced civilizations in the universe, says that the current equation predicts the existence of one technologically evolved planet per every 10 million star systems. The Drake Equation says that the number of such advanced civilizations is equal to the sum of the number of stars born per year in our galaxy, times the number that have planets, times the number with planets in a habitable zone, times the fraction of those that go on to develop life, times the fraction that develop intelligent life, times the fraction of such civilizations that reach a level of technological civilization capable of sending out signals, times the length of time they would be sending out detectable signals of life. The equation is widely discussed but, because it contains so many unknowns, it is more a conversation starter than a conversation finisher. Suffice it to say that the number of star systems observed so far is many, many times fewer than the 10 million Drake believes are needed to find a technologically advanced civilization. "We probably won't have something really scientifically useful to say for another twenty-five years—even if Hat Creek grows to three hundred fifty dishes," he said.

It is a daunting task. While the Milky Way alone has some 100 billion stars and untold billions of planets, the distances between them all are so enormous, the possible methods of communicating are so varied, and the cosmic timing has to be just right. Think about it: Some form of life has existed on Earth for more than 3.8 billion years, evolved Homo sapiens have roamed the world for 250,000 to 400,000 years, but humans who know how to produce radio waves to be detected by presumed other civilizations have been around only a little more than 100 years and humans searching for radio signals from afar only 50 years. Given the time scales involved, that's very close to nothing. Assuming for a minute that other technologically advanced civilizations have flourished, what's to say that happened just at the time when we can make contact with them?

I returned the next morning to wander the array again. This time it was sunny and Mount Shasta rose elegantly to the east. I could better see the surroundings, and the logic of SETI as well. It took a while to sink in as I made my way around, but astrobiology needs SETI, despite the skepticism, the feel of science fiction, that it inevitably brings. Astrobiology may well find signs of living or once-living extraterrestrial organisms in the not-too-distant future, and that would be a discovery of Copernican proportions. But what would really turn the world on its head is the discovery of intelligent life elsewhere. And SETI, for better or worse, is the pathway—actually, the shortcut—to a discovery of that magnitude.

• • •

With potentially so much at stake, the SETI field has come alive in recent years with alternatives to the tried-and-true technology of radio astronomy. The California physicist twins Gregory and James Benford, for instance, believe that SETI is listening in on the wrong radio channels, that alien civilizations would be more likely to send out tiny bursts of sound than the kind of broadband blasts the SETI Institute has been listening for. Their approach, James Benford said, "is more like Twitter and less like *War and Peace*."

Another ongoing effort involves searching for the potential existence of "Dyson spheres," hypothetical structures built by very advanced civilizations to shroud and thereby better utilize the energy from their sun. The logic and possible creation of these spheres was first put forward by physicist and mathematician Freeman Dyson. In a thought experiment in the late 1950s, Dyson concluded that because energy is the key to all advanced civilizations and the central star is the key to all planetary energy, it would make sense that any technologically sophisticated society would seek to maximize the energy available. This is especially true because all suns eventually lose their strength. In SETI terms, a Dyson sphere would give off strong signals that would establish the existence of an advanced civilization. Richard Carrigan, a particle physicist at the Fermi National Accelerator Laboratory in Illinois, and who is now active in SETI work, has been using data from an infrared satellite that covered 96 percent of the sky to search for stellar signatures that would resemble a Dyson sphere.

But the biggest, most consequential challenge to traditional SETI involves active searching for distant civilizations through sending out radio beams and laser pings, rather than remaining almost exclusively in a listening mode. An international protocol discourages "active" SETI (or METI, Messaging to Extraterrestrial Intelligence) but some have gone out anyway. The first came from the United States, or rather from the American-run Arecibo radio telescope in Puerto Rico. It was a much-debated digital message, featuring symbolic and coded illustrations of who we are and where we live, sent in the direction of a star system in the Hercules constellation about twenty-five thousand light-years away.

In 2008, the National Space Agency of Ukraine beamed a radio message in the direction of the star Gliese 581, which is known to have planets orbiting it. The next year, *Cosmos* magazine in Australia, with the support of the Australian government and NASA, solicited messages from around the world to beam in the same general vicinity, and some twenty-five thousand messages were collected from 195 nations. The 160-character text messages were transmitted from the Canberra Deep Space Communication Complex at Tidbinbilla, Australia. Both the Ukrainian and Australian transmissions will take about twenty years to reach their destination. But while these efforts were undertaken with benign intentions, they actually entered into a very controversial minefield. Some see active SETI as a logical next step, a shortcut to a shortcut. Others see it as dangerous, bordering on suicidal. The differences in approach are grounded in the most basic of human emotional responses: fear of the "other" versus a belief in the essential goodness, or at least the peacefulness, of the "other."

The proponents of active SETI assume that any civilization they would have the good fortune to contact would be harmless and interested in who we are and what we know. There's no evidence that this is true or false; it's just what they believe. They also hold that other technologically advanced civilizations would, by definition, be more advanced than us since we are just now reaching the stage where we can think about intergalactic communication. Perhaps they're waiting for us, and other civilizations like ours, to show we are here and capable and worthy of some form of contact. Doug Vakoch of the SETI Institute is one vocal advocate of sending out messages. He assumes that other and more advanced civilizations would not only be able to direct and tailor their intergalactic messages, but would almost certainly be able to decode messages from us much better than we could decode messages from them. Active SETI could speed up a possible "contact" in so many ways.

Vakoch believes his community is moving in the direction of active SETI, just as it is moving from radio listening to optical viewing. While freelance METI efforts are still frowned on—seen more as public relations than as science—more well-founded messages conceived by reputable groups or national space agencies are seriously considered and debated. METI advocates, he said, talk of the "zoo hypothesis," where humans would take the initiative and send out messages in all directions, to see who in the "galactic zoo" might respond. "Some people are concerned that alien life will come and eat us up but, well, it's not a concern of mine."

But it definitely is a worry for others. Opponents of active SETI assume intelligent extraterrestrial life, if it exists, will be warlike and intent on domination or destruction—rather like humans have often been as they overspread the Earth over the past two hundred thousand years. A sobering corollary to this view is that the number of technologically advanced civilizations in the universe may well be small because they tend to self-destruct, with the example of Earth as exhibit A. Our own shortsightedness about the stewardship of our planet and our inherent aggressiveness, in this view, will limit long-term survival. When the SETI community met in Valencia, Spain, as part of the 2006 International Astronautical Congress, the group voted against promoting active SETI, though it did not move to forbid it, either. The reasons why were numerous, including a desire to better focus the use of limited resources, but Harvard's Paul Horowitz captured an intellectual position that, oddly, tracks a key argument of the advocates of active SETI. "Statistically," he said, "it is extremely unlikely that our first contact with an ETI civilization will also be its first contact with an ETI civilization. Thus the advanced technology we detect will have experienced this type of encounter many times before."

In a 2010 documentary, none other than Stephen Hawking took concern about active SETI several steps further, warning that trying to actively contact possible intelligent life out there is way too risky and dangerous. Convinced that other higher life forms do exist in the galaxies, he doubted they would come in peace but rather would more likely raid Earth for its resources and then move on. As someone who also advocates in favor of terra-forming Mars as a sanctuary for Earthlings after we ruin the planet, there is a certain irony to Hawking's views.

"We only have to look at ourselves to see how intelligent life might develop into something we wouldn't want to meet," he wrote. "I imagine they might exist in massive ships, having used up all the resources from their home planet. Such advanced aliens would perhaps become nomads, looking to conquer and colonize whatever planets they can reach. If aliens ever visit us, I think the outcome would be much as when Christopher Columbus first landed in America, which didn't turn out very well for the Native Americans."

It all reminded me of the evening I spent with Shin-ya Narusawa after the SETI observation, a cold, clear night when a good number of astronomy enthusiasts came to Nishi-Harima to hear about Sazanka and to observe the sky. Narusawa had begun to get reactions from the public about his effort, and some of it was shrill. I watched him wince as a woman on the phone admonished him for letting potentially hostile aliens know where we are, with the assumption that they would then do us harm. Narusawa said that any SETI experiment results in calls like that—it just goes with the territory. That Narusawa and his colleagues were listening and watching rather than beaming out signals or messages was lost on the caller, as was the possibility that distant civilizations might actually be benign. But both the excitement and the fear about extraterrestrial life tend to be primal, a blank screen on which people project their own feelings and beliefs. Untold generations of humans have told stories about who or what might be out there, without having any evidence to guide them. At least now scientists are actually making an effort to search out and collect that evidence.

Standing on the observatory deck in Sayo, many of the guests wanted to see where the Sazanka team had pointed their instruments the nights before. What they saw was all a blur, even with the help of several smaller telescopes brought out for the occasion, but the people were fascinated just the same. They shivered as they waited their turns, and delighted in focusing on a planet or a star or a distant galaxy. Some lingered and seemed to be squinting hard, as if they might see something moving, some sign of life if they only looked hard enough. Just a little more effort—better instruments, better science, a little luck—and the world could change forever.

10 THE DAY AFTER FIRST CONTACT

Science has yet to discover extraterrestrial life, but there's a sense at the top of some of the world's dominant faiths that they must be prepared for the possibility: to defend the faith, to enhance the faith, to expand thinking about the role of humans in the universe, to make sure any confirmed extraterrestrial news of that sort is presented, and received, with calm and maybe even some awe. If finding extraterrestrial life would represent a coming full circle from the times and discoveries of Copernicus—who first persuasively demonstrated that the Earth circles the sun, and not the other way around—then religious leaders want to make sure their official views are ready. Copernicus, after all, famously refused to publish his revolutionary work until he was on his deathbed, for fear of ecclesiastical repercussions, and Galileo ran afoul of Church authorities for his astronomical findings and spent his last decade under house arrest.

But most damning of all, the Inquisition put to death the former monk, writer, and Renaissance philosopher Giordano Bruno for, among other heresies, his belief in a "plurality of worlds"—the existence of life on other celestial bodies. Bruno was burned at the stake in 1600 at Campo de' Fiori in Rome. A statue of him now dominates the square just a half mile from the Vatican where, in 2009, the Church's Pontifical Academy of Sciences held its first major conference on astrobiology. Convened on private Vatican grounds in the elegant Casina Pio IV, formerly the pope's villa, the gathering of prominent scientists and religious leaders discussed for four days the tenacious worlds of extremophiles, the burgeoning list of exoplanets, the discovery of methane on Mars, and how to read "biosignatures" throughout the cosmos.

"Astrobiology is a mature science that says very interesting things that could change the vision humanity has of itself. The church cannot be indifferent to that," is what Pierre Léna, a French astrophysicist and member of the Pontifical Academy, told me about the closed gathering.

The possibility of extraterrestrial life is not so much of an issue for Eastern religions, which speak explicitly of other inhabited worlds. Ancient Hindu texts describe innumerable universes inhabited by life forms both material and spiritual, as do early Buddhist holy books. Islam's Quran states, "All praise belongs to God, Lord of all the worlds," a phrase generally interpreted to mean the existence of many universal bodies and even multiple universes that may be inhabited. Some Western faiths also appear to make room for extraterrestrials: The Jewish Talmud refers to God "flying through eighteen thousand worlds," which later scholars have written implies that some or many are inhabited. Modern Christian groups including the Mormons also make a specific reference to extraterrestrial life—beings on other planets thought to be the same as humans or similar to them. It is with more traditional Christianity, where the coming of Christ to save sinners on Earth, and seemingly Earth alone, is so central, that astrobiology would seem to pose the biggest challenge. Some scientists are bracing for what they believe will be an inevitable and ugly conversation with leading conservative Christians, especially if the universe produces evidence of intelligent life.

"If there are beings elsewhere in the universe, then Christians, they're in this horrible bind," says Paul Davies, the physicist-astrobiologist and someone who has written extensively on the question. "They believe that God became incarnate in the form of Jesus Christ in order to save humankind, not dolphins or chimpanzees or little green men on other planets."

A more street-level view of the coming showdown comes from Cynthia Crysdale, a theology professor at the University of the South in Sewanee, Tennessee, and formerly of the Catholic University of America, during a NASA-supported workshop in 2006 on the implications of astrobiology. The problem, she said, is that the discovery of extraterrestrial life will push humans further still from the cosmic limelight. Copernicus and Galileo told us that the Earth was not the center of the universe, Darwin told us that we are the result of random mutation and the survival of the fittest, and now we're on the threshold of learning that life may well exist elsewhere. "This," she said, "won't go down lightly." In some conservative evangelical circles, it already hasn't. Gary Bates, the head of Creation Ministries in Atlanta, speaks regularly about the danger to Christianity of an acceptance of extraterrestrial life. "My theological perspective is that ET life would actually make a mockery of the very reason Christ came to die for our sins, for our redemption," he told me in connection with the Vatican conference. Bates believes that "the entire focus of creation is mankind on this Earth" and that intelligent, morally aware extraterrestrial life would undermine that view and belief in the incarnation, resurrection, and redemption drama so central to the faith. "It is a huge problem that many Christians have not really thought about." Bates, a sober, thoughtful, and anything but fire-and-brimstone speaker, traveled to Roswell, New Mexico, in 2009 to make similar warnings at a fundamentalist Christian counterconference on the officially designated anniversary of the supposed crashing of a UFO with "aliens" there in 1947.

But to assume an implacable response from the Christian corner may underestimate the faith's imagination and historical adaptability. If you listen to what some Catholic leaders, at least, have said on the subject of extraterrestrial life, the tone is generous and the argument even accommodating. In a front page story in the official Vatican newspaper, *L'Osservatore Romano,* José Gabriel Funes, a Jesuit astronomer from Argentina who is the director of the centuries-old Vatican Observatory, referred to our potential "extraterrestrial brothers" out in space and explained there was no theological reason to fear them.

Funes also heads the Vatican Advanced Technology Telescope observatory in the hills of Arizona outside Tucson, where Jesuit priests study the skies with a sophisticated 1.8-meter telescope. A Vatican observatory was initially founded in Rome in the eighteenth century to better calculate the arrival of holy days, which was done then through astronomy. The observatory was moved several times and eventually a second research center was established on Mount Graham in Safford, Arizona. Now the scientist-priests study galaxy formation, planetary sciences, meteorites, and the logic of the Big Bang. And from a vantage point far from Rome they clearly debate the big questions about what or who might inhabit those distant galaxies. For three decades they did so under the tutelage and guidance of Father George Coyne, a highly respected astronomer, historian, and advocate of science. "I have friends who pray that science will never discover or explain certain things," he once told *Wired* magazine. "I don't understand that. Nothing we learn about the universe threatens our faith. It only enriches it." Some say that his views, especially his strong critique of intelligent design thinking, led to his ouster as observatory director in 2006. Whether or not that is true, his successor seems to have taken in Coyne's teachings during his six years as an observatory researcher before becoming director. In his *L'Osservatore Romano* interview, Funes declared: "As a multiplicity of creatures exists on Earth, so there could be other beings, also intelligent, created by God. This does not conflict with our faith because we cannot put limits on the creative freedom of God."

That view was reprised, of all places, on *The Colbert Report,* the fake cable TV news show that often gets views through the back door that would never make it through the front. Following reports about the Vatican astrobiology conference in 2009, host Stephen Colbert invited one of the Vatican astronomers from Arizona, Brother Guy Consolmagno, onto the show. Brother Guy, as he wanted to be called, was trained as a physicist at MIT before becoming a priest and later earned a doctorate in planetary science from Arizona State University, so he's an astronomy professional as well as a devout Catholic. But he had the same welcoming embrace ready for any "extraterrestrial brothers." After all, he told Colbert, the Church has highlighted extraterrestrials for centuries in the form of angels. "The whole mythology of angels in the Christian and Jewish tradition shows that the Church, the people who wrote the Bible, were not afraid of other intelligent creatures who are also worshipping God."

Colbert wasn't buying it. "Why would the Catholic church do this?" he asked in both the mock and serious incredulity of a playacting but serious Catholic. "Doesn't this upset our place at the center of God's creation?" Consolmagno responded that God wanted us to learn about the universe as a way of knowing His work. Unimpressed, Colbert replied, "If we accept there is alien life on other planets, doesn't that totally blow Jesus out of the water? He came down and became man, not creature." Then came the clincher. Could Jesus, or Jesus-like saviors, be walking among and saving intelligent creatures on other faraway planets, too? Answered Consolmagno: "I'm not there, but it could be."

This extraterrestrial debate actually goes back in recorded fashion to the fourth-century B.C. Greeks. Epicurus, the philosopher of pleasure and pain, argued what became known as the "plurality of worlds" position, that life did exist on other celestial bodies. Aristotle, however, held a strong one-world view that did not allow for life beyond Earth and, adopted by Christianity, dominated the Western world well into the Renaissance. Books by Steven Dick, a retired historian at NASA, and Notre Dame University professor Michael Crowe detail the impressive constancy and heat of the debate over the centuries, as well as the elaborate deductions of men including Descartes, Immanuel Kant, and Benjamin Franklin as they tried to reach and defend positions with essentially no scientific data. Much of the debate involved whether or not the Christian worldview allowed for the possibility of a "plurality of worlds."

One of the most strident voices was that of founding father Thomas Paine, who, in *The Age of Reason,* wrote that a belief in Christianity and in what was then considered the enlightened acceptance of a plurality of worlds "cannot be held together in the same mind; and he who thinks that he believes in both has thought but little of either." In mid-nineteenth-century England, the man considered the greatest intellect of the time—geologist and moral philosopher William Whewell, master of Trinity College, Cambridge University—wrote first anonymously and then publicly that those who argued many planets were inhabited were wrong on the science, and also that central aspects of Christianity and the possible presence of extraterrestrial life *were* incompatible.

Whewell's defense of the Christian narrative against the "plurality of worlds" caused an intellectual firestorm that inspired some fifty articles and twenty books within several years. The biggest point of contention was not that belief in Christianity was undermining the evidence of science, as might be argued today. Rather, the plurality-of-worlds position, which was then taught in many schools, was believed by most men and women of science to be entirely consistent with the existence of an omnipresent God. What made Whewell's scientific and intellectual critics irate was that he was implying they had not thought through the implications of their convictions about the plurality of worlds and were therefore compromised Christians.

The implications of astrobiology hardly end with religion. A 2009 workshop on the possible societal challenges raised by astrobiology introduced a variety of others, many of them quite practical. In fact, that was the point of the two-day, NASA-sponsored meeting: to bring other disciplines into the discussion about how to respond to possible breakthroughs in extraterrestrial science. Those invited included forty-two professionals ranging from scientists to priests, from anthropologists and philosophers to ethicists. It was hardly a fringy group; many more men wore blue blazers than ponytails.

At the opening, SETI pioneer Frank Drake acknowledged that the task of the workshop, even the convening of the group, was on the far periphery of both science and astrobiology. Long an optimist about finding extraterrestrial life, however, Drake saw change coming. "When the time of a detection comes, this group will be among the most central in the world in dealing with the consequences, good and bad, of what just happened."

Margaret Race, the workshop organizer and a SETI Institute biologist, likened the astrobiological moment to the genetics and biotechnology revolutions of the 1990s, when scientists began to regularly move genes around and a layer of ethicists came in to set limits and create guidelines. "Scientists and engineers have been making the decisions about how to conduct missions and investigate space since the start of the space age," she told the group. "We now need others."

The meeting divided into groups based on areas of expertise, and over the two and a half days the list of issues to confront grew quite long. Some involved more familiar topics such as how to define life, how to look for life, and how to integrate potential extraterrestrial life into established religious and social views. But other, seldom broached questions came up that put astrobiology into a different light. For instance, heated discussions broke out about the moral responsibilities of humans to life found on other planets. And if a responsibility did exist, was it absolute, or did it slide up and down a scale with the simplicity or complexity of the life? Taking it a step further, some wondered whether extraterrestrial life could be patented and owned by future explorers. The workshop also heard discussion about terraforming, the process of modifying a planet (usually Mars) enough to create an atmosphere and then a biosphere that could support humans. Once the stuff of science fiction, it's now the subject of serious scientific and ethical discussion.

The modern religious, ethical, even philosophical debate about the current and future findings of modern astrobiology is in its infancy. But the purely secular debate about extraterrestrial life—the one that considers whether the discovery is possible at all—is fast losing ground to the march of science. The three primary viewpoints in that debate still have their advocates but here's why, at the finish of my own extended encounters with astrobiology, I have come to believe the science only points to one conclusion.

OPTION 1. We are alone in the universe and Earth is the only planet, moon, asteroid, comet, or undiscovered other body anywhere with life.

From the perspective of science and logic, this position is essentially a nonstarter. We can search for life beyond Earth for centuries and not find it, but that doesn't mean it isn't there. It just means we didn't find it. As philosopher and astrobiology thinker Carol Cleland put it, this is not a case where the absence of evidence is evidence of absence. From a logical perspective, she said, sometimes an absence of evidence can and does mean something doesn't exist. If you go to Africa on the hunch that the brontosaurus still roams the land and you don't find any, then the absence of evidence is compelling. But that isn't the case with our efforts so far to find life of any sort beyond Earth. Astrobiology is a huge and dynamic endeavor and it has, in its modern incarnation, only just begun. Especially given the vastness of the canvas and the discoveries of the past decade, I don't see how anyone can reasonably come to the conclusion or make the claim that Earth and only Earth has life.

OPTION 2. Only Earth has complex life.

Sure, microbes might exist in nooks and crannies, deep underground or in liquid or icy water, but too many things have to go right for a life-form to evolve beyond a basically single-cell state. This is the argument of the influential book *Rare Earth,* by paleontologist Peter Ward and astronomer Donald Brownlee. While microbial life is probably widespread across the cosmos, they write, it takes enormous good fortune for life to evolve into more complex forms because of the innumerable ways that extinction—to the point of planetary sterilization—can be inflicted. Even the Earth has had many close calls that could have snuffed out all life; the fact that evolution continued, they argue, is something of an improbable fluke. The argument has its followers, but I find the conclusion unconvincing.

First, having "life" come together from nonlife is generally viewed as the hard part, the step that scientists don't understand and can't replicate. Even when future headlines read that "life" has been created in a lab, the claim will be a substantial exaggeration for years to come since the "life" created will be made from molecules and compounds manufactured by humans. What synthetic biologists are doing is remarkable, but nobody is close to understanding how life can be formed from nonliving parts that haven't been previously engineered or changed by scientists. That's why the cartoon resonates of a professor at the blackboard diagramming the origin of life in exquisite but incomplete detail. The blackboard equations and processes lead to life only because toward the end of the intricate formula, a balloon appears that matter-of-factly declares, "Miracle Happens." I'm not saying that life needs an otherworldly miracle to begin, or that science isn't making progress in refining possible scenarios. Rather, the evidence needed to scientifically know what happened on early Earth is gone, erased by 4.5 billion years of turbulent geology and changing atmospheres, and some 3.8 billion years of evolution.

So starting life is, by most accounts, the biggest deal. Keeping it going and allowing it to evolve is also a big deal, and is where the "Rare Earthers" find disqualifying obstacles. So many stellar, solar, and planetary factors have to be just so for evolution to produce complex life that the chances of it happening frequently are slim to none, they argue. They might be right. But this presumed specialness of the cosmic dynamics that allowed Earth and perhaps only Earth to be stable long enough for single-cell life to evolve into something more complex sounds familiar, suspicious even. Earth was special, too, when people believed it was the center of the universe, and it is unique today to those who reject the logic and power of evolution and natural selection as a pathway from simplicity to complexity. Ward and Brownlee certainly don't share those beliefs, but they are awfully impatient. Given the fact that we have known for sure only since 1995 that planets exist outside our solar system, and given the fact that talented and engaged scientists strongly believe that innumerable stars have habitable zones for the planets that orbit them, and given the fact that the laws of physics and chemistry appear to work throughout the universe as they do here, it seems very premature to say that complex life evolved on Earth alone. Especially so, now that respected astronomers like Geoff Marcy and Andrew Howard have extrapolated from their planet-hunting data that tens of billions of Earth-sized planets exist in the Milky Way alone.

What's more, as any microbiologist will tell you, single-cell organisms come in species from the rudimentary to the extremely complex, they contain the same kind of DNA-based genetic material that organizes the rest of "higher" life, and they ruled the Earth alone for most—more than 75 percent—of the time that Earth has supported life. So if microbes are common across the universe, then it strains credulity to conclude that evolution everywhere else stops with them. Certainly, single-cell life on other distant planets can be easily snuffed out by an incoming asteroid, a fading sun, or any number of other catastrophes. But if astrobiology has proven nothing else, it has unequivocally shown that life, once started, is tenacious beyond imagination, and that it can adapt to the most extreme circumstances.

OPTION 3. Life exists beyond Earth and, in some instances, has become complex and most likely includes what we would consider intelligence.

Not surprisingly, this is what many and probably most scientists involved in astrobiology consider the most likely scenario. An objection often raised to this conclusion is that if extraterrestrial life existed on other planets or moons, we would have already found it. Similarly, people ask, if intelligent alien life exists, why hasn't it come to Earth and made itself known? Variations on these questions were famously posed by nuclear physicist Enrico Fermi in 1950: If the perceived probability of the existence of extraterrestrial civilizations is so high, he asked, then why is there no evidence for, or contact with, any life-forms beyond Earth? Fermi's paradox is not an insignificant obstacle. It was posited around the time that UFO reports—perhaps the layman's answer to the paradox—became the rage. Absent a good answer, some people turned to a bad one. But other answers exist, and many again focus on the vastness of time and space.

Humans have had the capacity to actually search for extraterrestrial life in a technologically advanced way for only about fifty years. We on

Earth have produced radio waves and other far-traveling signatures of intelligent activity for less than 150 years. In a universe estimated to be 13.8 billion years old, and which has had stars, solar systems, and trillions of planets for a good portion of that time, 150 years is not even a blink of the eye. What's more, the stars closest to our sun are the Alpha Centauri system, about 4.25 light-years away. That's about 30 trillion miles. No planets have actually been found in Alpha Centauri, but let's assume for a second they exist and house intelligent life. Traveling at the speed of light, a beamed message going back and forth to our solar system would take 8.5 years.

It's true that the biosignatures of atmospheric oxygen and complex hydrocarbons have been around Earth's atmosphere much longer, and technologically advanced civilizations from beyond could have detected them and known what they meant. Logic says that is correct, but doesn't offer much more on the question. But science, and especially astrobiology, is grounded in this different truth: An absence of evidence is not evidence of absence—unless you're looking for living brontosauri or unicorns in the ballfield. It may be meaningful or frustrating to some that extraterrestrial life remains unfound or unconfirmed but, really, the search has just begun. When Fermi posed his paradox, it actually *hadn't* even begun. Step back a bit, and this picture emerges of what has been discovered: The universe is now ruled by what appears to be a singular set of laws of physics and chemistry, organizing principles that led to the formation of single-cell microbes on one planet at least and, through evolution, led to ragweed, cats, and us. The universe is now known to be swimming in carbon, amino acids, nitrogen, hydrocarbons, water, and energy from starlight—the bottom-line essentials for life as we know it. And life on Earth has shown itself to be extraordinarily tenacious and capable of surviving catastrophes from asteroid hits to the glacial times of what is sometimes called Snowball Earth. If life started on any of those countless other planets we know to be out there, why wouldn't it have the same tenacity? The universe, it seems, has the ingredients and structure for quite a cosmic menagerie.

Yes, spacecraft have landed on or flown past most of the planets and some of the moons of our solar system, and no life has yet been confirmed. What's more, we've found that conditions out there can get awfully cold, awfully hot, awfully dry, and awash in killing cosmic radiation. But many planetary scientists hold that the possibility of some current or former life on Mars, Europa, or Enceladus in particular remains very much intact, and our ability to identify and confirm its presence is in its infancy. Any discovery of a second or third genesis in our own pedestrian solar system would pretty much have to mean that life is a commonplace across the universe. So the real long-term action is those exoplanets, the ones we'll be examining with occulters and sunshades and spectrometers on the lookout for signatures of life. Doesn't it seem unlikely that none—zero—of the trillions and trillions of planets now reasonably presumed to exist beyond our solar system have the ingredients and conditions needed to cobble together life, and the stability needed to allow life to evolve and grow more complex? Doesn't it seem more likely that we will, in the near or further future, make our first contact with undoubtedly extraterrestrial life, and that it will be a day our world long remembers as the harbinger of a new frontier in a dramatically changed cosmos?

ACKNOWLEDGMENTS

My indebtedness to the insights, skills, and kindnesses of others in creating this book is perhaps best told through chronology. It was due to a series of hardly inevitable occurrences that the project was started and through the generosity of others' spirit that it grew and thrived.

It was my *Washington Post* colleague Shankar Vedantam who got the ball rolling. He had participated in the Templeton-Cambridge Journalism Fellowship in Science and Religion at Cambridge University, and for several years he suggested I, too, might find it valuable and enjoyable. My resistance broke down in 2007, and I'll be forever grateful to Shankar for his persistence. My month at Cambridge in 2008 was spent listening to talks by prominent and often compelling speakers from many disciplines, and then was followed by my writing and presenting a paper on the topic I had selected—astrobiology and its implications for religion. The Cambridge program was expertly run with great hospitality by Fraser N. Watts, Director of the Psychology and Religion Research Group at Cambridge University and an Anglican priest; Julia Vitullo-Martin, director of the Center for Urban Innovation in New York; and Sir Brian Heap, a research associate at the University of Cambridge. Their help and encouragement was essential, and the fellowship provided an ideology-free setting for learning.

Part of the program involved having one's work published, and, after an expert edit by my *Washington Post* editor Nils Bruzelius, my story about astrobiology did appear in the newspaper. Even before that, however, I was contacted by Washington literary agent Gail Ross, who wanted to know if I was interested in writing a book based on my Templeton work. I was again hesitant, but, after meeting with her and editorial director Howard Yoon, I saw how my germ of an idea could grow. The talents, insights, and enthusiasm they brought to that initial conversation still make me smile.

My good fortune continued when Priscilla Painton, executive editor of Simon & Schuster, concluded that my book proposal was something that she and her imprint wanted. What followed was two years of highly productive work with an editor who was not only superb at all aspects of editing, but was always available to give direction and advice, and was imaginative, savvy, and fun. Her skills, both personal and professional, were absolutely essential to giving the book its shape and reach. Her assistant, Michael Szczerban, was similarly a talented and professional pleasure to work with.

My travels kept me on the road constantly and required a not-insignificant investment of our family funds, but my wonderful wife, Lynn Litterine, encouraged me all the way. She also listened with care to my adventures and scientific insights (and misunderstandings) and gave much-valued feedback as I described both the science and the scientists I was meeting. A partner could not ask for more. I'm indebted as well to my father, Irving Kaufman, for support that was intellectual, moral, and financial, and wish that my mother, Mabel Kaufman, had been alive to watch the process unfold. Our two sons, David and John Litterine-Kaufman, helped provide the emotional richness that makes a project like this possible. I especially enjoyed talking with John (a veterinarian in training) about science, and David and his wife Elizabeth Nolte were a joy to be with in Istanbul—where they were living for a year and where Lynn and I stopped on my round-the-world reporting run.

Other friends, colleagues, and scientists who helped in many ways not always visible were Kathy Sawyer, Frances Sellers, Nils Bruzelius, Phillip Bennett, Vanessa Gezari, Rob Stein, Adele Nakayama, Yoko Hisakata, Liz Gulliford, Jonathan Trent, Radu Popa, Jonathan Lunine, Farid Salama, John Rummel, Ernan McMullin, Connie Bertka, Gary Rosen, Shawn Doyle,

Amanda Achberger, Derek Litthauer, Seth Shostak, Laura Ventura, Marc and Deborah Taylor, Emily Yoffe, and Peter Perl.

As someone who writes about science but is not trained as a scientist, I had the sometimes daunting task of trying to understand complex issues in a wide range of scientific disciplines and languages. Most of the subjects of *First Contact* agreed to read parts of the book behind me and correct any misunderstandings. Errors that may remain are entirely my own.

NOTE ON SOURCES

I went into my reporting of *First Contact* with more than three decades of journalism and writing experience but limited knowledge of the many scientific disciplines that make up astrobiology. As a result, I was utterly dependent on the time and gracious teaching of the scores of researchers who let me into their labs and into their lives. Almost all were accustomed to speaking about their research primarily with their scientific peers, yet my early lack of knowledge about their fields and their terminology required me to ask many, many questions. I never once felt that the scientists resented my endless inquiries; instead, I found them eager to talk until their work was understood in a way that it could be accurately explained to scientists and nonscientists alike. I will be forever grateful for the education I received.

While I spoke with scores of women and men about their work, I am very much aware of the many other important and talented scientists working in astrobiology whom I did not interview or get to know. This book describes the work of a limited number of researchers, and I am sure all would say they could not have achieved the breakthroughs they did (or will) without the prior work of many others and the current work of many colleagues.

As mentioned in the book's opening chapter, my interest in astrobiology was excited by one of its foremost practitioners—theoretical physicist and exoplanet expert Sara Seager of MIT. Her knowledge and enthusiasm helped both direct and inform me. Edward Weiler, associate administrator for NASA's Science Directorate, also gave me the broad overview of astrobiology that I needed to get started, and the confidence that the enterprise was headed somewhere very important. Both Mary Voytek, NASA's chief scientist for astrobiology, and Carl Pilcher, the director of the NASA Astrobiology Institute, played similar foundational roles, as did Fraser Watts and Lord Martin Rees of Cambridge University. Linda Billings, research professor, School of Media and Public Affairs of George Washington University, was instrumental as well in setting me in the right directions, as was Arthur Charo of the National Research Council. The textbook *Planets and Life: The Emerging Science of Astrobiology*, by Woodruff T. Sullivan III and John A. Baross, provided invaluable background understanding. And of course, every writer about astrobiology is in some way influenced by the late Carl Sagan.

The opening chapter is a compilation and interpretation of the knowledge I gleaned from my scores of interviews, in addition to readings of books and scientific papers listed in the bibliography. But a number of people stand out as especially informative and helpful in identifying major themes: Sara Seager; Steven Benner of the Foundation for Applied Molecular Evolution in Gainesville, Florida; Carol Cleland of the University of Colorado, Boulder; Tullis Onstott of Princeton University; Michael Mumma of NASA's Goddard Space Flight Center; Steven J. Dick, former official historian for NASA; David McKay of NASA's Johnson Space Center; Paul Butler of the Carnegie Institution of Washington; Pan Conrad of the Goddard Space Flight Center; and Christopher McKay of NASA's Ames Research Center.

The chapter on extremophiles was anchored in the pioneering work of scientists such as John Baross of the University of Washington and John Priscu of Montana State University, but relied heavily on time spent with Tullis Onstott, Brent Christner of Louisiana State University, Gaetan Borgonie of the University of Ghent, and Lisa Pratt of Indiana University. My journey to the mines of South Africa was accomplished only with the assistance of Esta van Heerden and Derek Litthauer of the University of the Free State in Bloemfontein, South Africa, who both took me along with them and taught me along the way. Peter Doran of the University of Illinois in Chicago and Bill Stone of Stone Aerospace in Austin, Texas, explained the science and technology involved in the Lake Bonney expedition in Antarctica. In addition, the 2010 American Geophysical Union's Chapman Conference on the Exploration and Study of Antarctic Subglacial Aquatic Environments gave me a three-day tutorial on the improbable yet central world of microbes living in ice.

Exploring the definitions of life was inspired by and aided substantially by the work of Carol Cleland and Christopher Chyba of Princeton University, and especially by the patient teaching of Cleland. In addition, Gilbert Levin, now president of Spherix and formerly a principal investigator for the NASA Viking mission, shared his information and time, as did NASA's Michael Meyer, chief scientist for Mars missions. Penelope Boston of New Mexico Tech, Kimberly Kuhlman of the Planetary Science Institute, and Ferran Garcia-Pichel of Arizona State University explained and showed me the world of desert varnish. Steve Benner and his Foundation for Applied Molecular Evolution colleague Matt Carrigan explained synthetic biology and allowed me several days in their lab, and Gerald Joyce and Tracey Lincoln described their groundbreaking work in creating self-replicating RNA at the Scripps Research Institute.

Regarding the Miller-Urey experiment and its decades of influence, I am most indebted to Danny Glavin and Jason Dworkin of NASA's Goddard Space Flight Center for their generous sharing of time and our trip together to the National Museum of Natural History in Washington to discuss the Murchison meteorite. They also gave me insights into the phenomenon of chirality, as did Sandra Pizzarello of Arizona State University. Jeffrey Bada of the Scripps Oceanographic Institution, Antonio Lazcano of the National Autonomous University of Mexico, and Rafael Navarro-González of the university spent substantial time teaching me, as did Pascale Ehrenfreund of the University of Leiden and George Washington University. My journey to Alaska was greatly aided by lightning experts Ronald Thomas of New Mexico Tech and his colleague Sonje Behnke, and volcanologist

Steven McNutt of the Alaska Volcano Observatory. McNutt arranged for our remarkable trip to the mouth of the Mount Redoubt volcano.

Michael Mumma of Goddard Space Flight Center not only explained the complex world of spectroscopy and how he and his colleagues detected methane on Mars, but he also helped secure an invitation for me to meet him at the European Southern Observatory at Paranal, Chile. He and his colleague, Geronimo Villanueva, allowed me to join them for two nights of observing, which was both fascinating and illuminating. Laura Ventura of the ESO also helped instruct me on some of the mechanics of the telescope and astrophysics of the work. Edward Weiler of NASA described the importance of Mumma's work in reaching an agreement with the European Space Agency for two joint missions to Mars. For the section on additional research into possible life on Mars, I relied on interviews with Paul Mahaffy of NASA's Goddard Space Center, Steven Squyres of Cornell University, and the rich literature on the long-ago presence of water on the Martian surface, the former presence of a magnetic field, and the mineralogy of the planet. Most important, I learned firsthand from NASA's Pan Conrad at Death Valley, the Mojave Desert, and White Mountain Peak about how to analyze habitats in extreme areas. As a member of NASA's Mars Science Laboratory science team, which will analyze Mars for organic material and habitats potentially suitable for life, she gave me insights into how the mission will unfold. In addition, numerous presentations at the Astrobiology Science Conference of 2010 in League City, Texas, about the Martian past helped inform my reporting.

In navigating the difficult shoals of the controversies about published research asserting that extraterrestrial life had been detected in Martian and other meteorites and by the Viking Mars lander in 1976, I spent substantial time with Richard Hoover of NASA's Marshall Space Center, David McKay of NASA's Johnson Space Center, and Gil Levin. All three explained in great detail how and why they are convinced they did discover extraterrestrial life. Contrary views were provided by Andrew Steele of the Carnegie Institution of Washington, NASA's Mary Voytek and Michael Meyer, Allan Treiman of the Lunar and Planetary Institute, and Rafael Navarro-González. Kathy Sawyer's *The Rock from Mars* gave deeply reported and highly respected insights into the ALH 84001 Martian meteorite debate, and Henry Cooper's *The Search for Life on Mars* did the same for the Viking Labeled Release controversy. Presentations at the 2010 Astrobiology Science Conference (AbSciCon) by Kathie Thomas-Keprta and Everett K. Gibson helped explain the current status of research on the famous meteorite.

The science of planet-hunting was illuminated and made fascinating primarily by Paul Butler, who invited me to spend time with him at the Anglo-Australian Observatory in Coonabarabran, Australia. His patient descriptions of the complex science helped me enormously. The paper published by Butler and Steve Vogt of the University of California at Santa Cruz on the discovery of the first exoplanet in a "habitable zone" provided the exclamation point at the end of the planet-hunting sentence. The literature on exoplanet discoveries is also rich, and I found James Kasting's *How to Find a Habitable Planet* to be useful. Webster Cash of the University of Colorado, Boulder, and Remi Soummer of the Space Telescope Institute in Baltimore explained the science of occulters after I was first introduced to it by Sara Seager. The complex history of NASA's efforts to develop a Terrestrial Planet Finder mission, which would explore for potentially habitable planets, was the subject of several presentations at AbSciCon 2010 as well.

My introduction to the issue of the "fine-tuning" of the universe occurred at Cambridge University, and came from Lord Martin Rees, history of science professor and minister Fraser Watts, and former physics professor and minister John Polkinghorne. Theoretical cosmologist Paul Davies of Arizona State University helped me in numerous ways in both interviews and his several books on the subject. Lee Smolin of the Perimeter Institute of Theoretical Physics taught me about cosmic natural selection, as did Louis Crane of Kansas State University. George Ellis of the University of Cape Town provided a religious interpretation of fine-tuning.

On the subject of the shadow biosphere and arsenic-based life in Mono

Lake, I relied on information from NASA Astrobiology Institute fellow Felisa Wolfe-Simon, her colleague at the U.S. Geological Survey, Ron Oremland, Mary Voytek and microbiologist Rosie Redfield at the University of British Columbia. Reporting by Carl Zimmer in *Slate* and Dennis Overbye at *The New York Times* was also informative. In addition, I was initially introduced to the subject of the possible shadow biosphere by Paul Davies and Carol Cleland, who both shared insights into their pioneering thinking.

The literature on SETI is extensive, but firsthand reporting is not. That is why I traveled to both the Allen Array at Hat Creek, California, to see the new SETI radio telescopes, and to the Nishi-Harima Observatory in western Japan to witness a nationwide SETI observation organized by observatory director Shin-ya Narusawa, who provided much-appreciated information and insight. Frank Drake, Jill Tarter, and Douglas Vakoch of the SETI Institute spoke with me at length about current SETI projects and history. Rick Forester is the chief scientist at Hat Creek, and he also provided invaluable information. I was introduced to the controversial and fascinating subject of "active SETI" by Vakoch and listened to several presentations about it at the AbSciCon 2010 gathering.

The history of the extraterrestrial debate, and especially its religious ramifications, is well documented by NASA's Steven Dick and the University of Notre Dame's Michael Crowe. I am indebted to Father José Gabriel Funes, director of the Vatican Observatory, for his help regarding the 2009 Vatican conference on astrobiology, and to Notre Dame University professor Ingrid Rowland for her work on the martyred Giordano Bruno. Fraser Watts of Cambridge University also helped put issues in perspective. Margaret Race of SETI organized the NASA-sponsored 2009 conference on the implications of astrobiology, and I appreciate her willingness to have me participate and share much information. My final conclusions in chapter 10 were largely the result of my own thinking, but they were helped by discussions with Carol Cleland, Geoffrey Marcy, Paul Davies, Ed Weiler, and Sara Seager, as well as reading the works of Peter Ward, Alan Boss, and Martin Rees.

BIBLIOGRAPHY

BOOKS

Benner, Steven. *Life, the Universe, and the Scientific Method.* Gainesville, Fla.: Ffame Press, 2008.

Bertka, Connie, Nancy Roth, and Matthew Shindell. *Workshop Report: Philosophical, Ethical, and Theological Implications of Astrobiology.* American Association for the Advancement of Science, 2007.

Boss, Alan. *The Crowded Universe: The Search for Living Planets.* New York: Basic Books, 2009.

Chaikin, Andrew. *A Passion for Mars: Intrepid Explorers of the Red Planet.* New York: Harry N. Abrams, 2008.

Cooper, Henry S. F. *The Search for Life on Mars: Evolution of an Idea.* New York: Holt, Rinehart & Winston, 1980.

Crowe, Michael J. *The Extraterrestrial Life Debate, 1750–1900.* Mineola, N.Y.: Dover, 1999.

Davies, Paul. *The Eerie Silence: Renewing Our Search for Alien Intelligence.* Boston: Houghton Mifflin Harcourt, 2010.

———. *The Goldilocks Enigma: Why Is the Universe Just Right for Life?* Mariner Books, 2006.

De Duve, Christian. *Singularities: Landmarks on the Pathway of Life.* New York: Cambridge University Press, 2005.

Dick, Steven J. *Plurality of Worlds.* New York: Cambridge University Press, 1982.

Dick, Steven J., ed. *Many Worlds: The New Universe, Extraterrestrial Life and the Theological Implications.* Radnor, Penn.: Templeton Foundation Press, 2000.

Hazen, Robert M. *Genesis: The Scientific Quest for Life's Origins.* Washington, D.C.: Joseph Henry, 2005.

Impey, Chris. *The Living Cosmos: Our Search for Life in the Universe.* New York: Random House, 2007.

Popa, Radu. *Between Necessity and Probability: Searching for the Definition and Origin of Life.* New York: Springer-Verlag, 2004.

Rees, Martin. *Just Six Numbers: The Deep Forces That Shape the Universe.* New York: Basic Books, 2000.

Rowland, Ingrid. *Giordano Bruno: Philosopher and Heretic.* New York: Farrar, Straus & Giroux, 2008.

Sawyer, Kathy. *The Rock from Mars: A Detective Story on Two Planets.* New York: Random House, 2006.

Seager, Sara. *Is There Life Out There: The Search for Habitable Exoplanets.* E-book published by author, www.saraseager.com, 2010.

Shostak, Seth. *Confessions of an Alien Hunter: A Scientist's Search for Extraterrestrial Intelligence.* Washington, D.C.: National Geographic Books, 2009.

Smolin, Lee. *The Life of the Cosmos.* New York: Oxford University Press, 1997.

Sullivan, Woodruff T., and John Baross, eds. *Planets and Life: The Emerging Science of Astrobiology.* New York: Cambridge University Press, 2007.

Ward, Peter, and Donald Brownlee. *Rare Earth: Why Complex Life Is Uncommon in the Universe.* New York: Copernicus, 2004.

PAPERS

Bada, Jeffrey. "How life began on Earth: A status report." *Earth and Planetary Science Letters,* 2004.

Bada, Jeffrey, Antonio Lazcano, et al. "The Miller Volcanic Discharge Experiment." *Science,* 2008.

Baross, John, and Jody Deming. "Growth of 'black smoker' bacteria at temperatures of at least 250°C." *Nature,* 1983.

Benford, James, et al. "Searching for Cost-Optimized Interstellar Beacons." AbSciCon conference, 2010.

Biemann, Klaus. "On the ability of the Viking gas chromatograph–mass spectrometer to detect organic matter." *Proceedings of the National Academy of Sciences,* 2007.

Boston, P. J., M. N. Spilde, D. E. Northup, et al. "Cave biosignature suites: Microbes, minerals and Mars." *Astrobiology Journal,* 2001.

———. "Evaporites in caves: A new paradigm for understanding microbial/mineral cave deposits." *Astrobiology,* 2005.

Bulat, S. A. "Cell Concentrations of Microorganisms in Glacial and Lake Ice of the Vostok Ice Core, East Antarctica." *Microbiology,* 2009.

Butler, R. P. "Extrasolar planets and the implications for life." In W. Sullivan and J. Baross, eds., *Planets and Life: The Emerging Science of Astrobiology.* Cambridge University Press, 2006.

Christner, Brent, et al. "Bacteria in subglacial environments." In *Psychrophiles: From Biodiversity to Biotechnology,* Springer, 2008.

———. "Geographic, seasonal, and precipitation chemistry influence on the abundance and activity of biological ice nucleators in rain and snow." *Proceedings of the National Academy of Sciences,* 2008.

Christner, Brent, John Priscu, et al. "Limnological conditions in Subglacial Lake Vostok, Antarctica." *Limnology and Oceanography,* 2006.

Cleland, Carol. "Epistemological issues in the study of microbial life: Alternate terran biospheres?" *Studies of History and Philosophy & Biological and Biomedical Science,* 2007.

———. "Understanding the nature of life: A matter of definition or theory?" In J. Seckbach, ed., *Life As We Know It.* Dordrecht: Springer, 2006.

Cleland, Carol, and Christopher Chyba. "Does 'life' have a definition?" In Sullivan and Baross, eds., *Planets and Life: The Emerging Science of Astrobiology.* New York: Cambridge University Press, 2007.

Cronin, John, and Sandra Pizzarello. "Enantiomeric excesses in meteoritic amino acids." *Science,* 1997.

Cunningham, K. I., D. E. Northup, et al. "Bacteria, Fungi and Biokarst in Lechuguilla Cave, Carlsbad Caverns National Park, New Mexico." *Environmental Geology,* 1995.

Davies, Paul. "Searching for an alternative form of life on Earth." *Proceedings of SPIE,* 2007.

Davies, Paul, et al. "Signatures of a Shadow Biosphere." *Astrobiology,* 2009.

Deming, Jody, Frank Drake, and Sara Seager. "Detection of infrared radiation from an extrasolar planet." *Nature,* 2005.

Drake, Frank, et al. "The New Telescope/Photometer Optical SETI Project of SETI Institute and the Lick Observatory." AbSciCon conference, 2010.

Ehrenfreund, P., and S. B. Charnley. "Organic molecules in the interstellar medium, comets, and meteorites: A voyage from dark clouds to the early Earth." *Astrophysics,* 2007.

Engel, Michael. "Partial racemization of amino acids in meteorites: Implications for their possible modes of origin." *Proceedings of SPIE,* 2009.

Garcia-Pichel, F. "Microbes and the search for life beyond Earth." *Fundamentals and Challenges of Astrobiology,* 2005.

Glavin, Daniel P., et al. "Amino acid analyses of Antarctic CM2 meteorites using liquid chromatography—time of flight—mass spectrometry." *Meteoritics and Planetary Science,* 2006.

———. "Extraterrestrial Amino Acids in the Almahata Sitta Meteorite." *Meteorites and Planetary Sciences,* 2010.

Glavin, Daniel P., and Jason Dworkin. "Enrichment of the amino acid L-isovaline by acqueous alternation on Ci and CM meteorite parent bodies." *Proceedings of the National Academy of Sciences,* 2009.

Hoover, Richard. "Chiral biomarkers and microfossils in carbonaceous meteorites." *Proceedings of SPIE,* 2010.

———. "Life in ice: Implications in astrobiology." *Proceedings of SPIE,* 2009.

———. "Microfossils in carbonaceous meteorites." *Proceedings of SPIE,* 2009.

———. "Microfossils of cynobacteria in carbonaceous meteorites." *Proceedings of SPIE,* 2007.

Howard, Andrew, and Geoffrey Marcy. "The Occurrence and Mass Distribution of Close-in Super-Earths, Neptunes, and Jupiters." *Science,* 2010.

Klussman, Martin, et al. "Thermodynamic control of asymetric amplification in amino acid catalysis." *Nature,* 2006.

Krehbiel, P. R., et al. "Observations of the Electrical Activity of the Redoubt Volcano in Alaska." American Geophysical Union, 2009.

Kuhlman, Kimberly, et al. "Enumeration, isolation, and characterization of ultraviolet resistant bacteria from rock varnish in the Whipple Mountains." *Icarus,* 2005.

———. "Evidence of a microbial community associated with rock varnish at Yungay, Atacama Desert, Chile." *Journal of Geophysical Research,* 2008.

Levin, Gilbert. "Extant Life on Mars: Resolving the Issues." *Cosmology,* 2010.

———. "Methane and Life on Mars." *Proceedings of SPIE,* 2009.

Levin, Gilbert, et al. "Viking Mars Label Release Results." *Nature,* 1979.

Lincoln, Tracey, and Gerald Joyce. "Self-sustained replication of an RNA enzyme." *Science,* 2009.

Martins, Zita, and Pascale Ehrenfreund. "Extraterrestrial nucleobases in the Murchison meteorite." *Earth and Planetary Science Letters,* 2008.

McKay, David S., et al. "Life on Mars: New Evidence from Martian Meteorites." *Proceedings of SPIE.* 2009.

———. "Search for Past Life on Mars: Possible Relic Biogenic Activity in Martian Meteorite ALH84001." *Science,* 1996.

McNutt, Stephen, and Earle Williams. "Volcanic lightning: Global observations and constraints on source mechanisms." Springer-Verlag, 2010.

Mielke, R. E., M. J. Russell, et al. "Design, Fabrication and Test of a Hydrothermal Reactor for Origin of Life Experiments." *Astrobiology,* 2010.

Mikucki, Jill, et al. "A contemporary microbially maintained subglacial ferrous 'ocean.'" *Science,* 2009.

Miller, Stanley. "A production of amino acids under possible primitive Earth conditions." *Science,* 1953.

Miller-Ricci, Eliza, and Sara Seager. "The atmospheric signatures of super-Earths: How to distinguish between Hydrogen-rich and Hydrogen-poor atmospheres." *Astrophysical Journal,* 2009.

Mumma, Michael, et al. "Strong Release of Methane on Mars in Northern Summer 2003." *Science,* 2009.

Nagy, B., G. Claus, and D. J. Hennessy. "Organic particles embedded in minerals in the Orgueil and Ivuna carbonaceous chondrites." *Nature,* 1962.

Nagy, Bartholomew. "The possibility of extraterrestrial life: Ultra-microchemical analyses and electro-microscopic studies of microstructures in carbonaceous meteorites." *Review of Paleobotany and Palynology,* 1967.

Narusawa, Shin-ya. "Project SAZANKA: The Multi-site and Multi-frequency Simultaneous SETI Observation in Japan." AbSciCon conference, 2010.

National Research Council. "The limits of organic life in planetary systems." National Academies Press, 2007.

Navarro-González, Rafael, et al. "The limitations on organic detection in Mars-like soils by thermal volatilization–gas chromatography–MS and their implications for the Viking results." *Proceedings of the National Academy of Sciences,* 2006.

———. "Mars-like soils in the Atacama desert, Chile, and the dry limit of microbial life." *Science,* 2003.

Navarro-González, Rafael, Christopher McKay, et al. "Reanalysis of the Viking results suggests perchlorate and organics at mid-latitudes on Mars." *Journal of Geophysical Research,* 2010.

Onstott, Tullis, et al. "The Origin and Age of Biogeochemical Trends in Deep Fracture Water of the Witwatersrand Basin, South Africa." *Geomicrobiology Journal,* 2006.

Onstott, Tullis, Dylan Chivian, et al. "Environmental Genomics Reveals a Single-Species Ecosystem Deep Within Earth." *Science,* 2008.

Perry, Randall, et al. "Baking black opal in the desert sun: The importance of silica in desert varnish." *Geology,* 2006.

Pizzarello, Sandra. "The Chemistry of Life's Origin: A Carbonaceous Meteorite Perspective." *Accounts of Chemical Research,* 2006.

———. "Molecular asymmetry in extraterrestrial chemistry: Insights from a pristine meteorite." *Proceedings of the National Academy of Sciences,* 2008.

Plaut, Jeffrey, et al. "Radar evidence for ice in lobate debris aprons in the mid-northern latitudes of Mars." *Geophysical Research Letters,* 2008.

Raymond, James, and Brent Christner. "An ice-adapted bacterium from the Vostok ice core." *Extremophiles,* 2008.

Renno, Nilton, et al. "The discovery of liquid water on Mars and its implications for astrobiology." *Proceedings of SPIE,* 2009.

Ricardo, A., M. Carrigan, and S. Benner. "Borate Minerals Stabilize Ribose." *Science,* 2004.

Rozanov, Alexi. "Bacterial Paleontology." *Perspectives in Astrobiology,* 2005.

Russell, M. J., et al. "Serpentinization and its contribution to the energy for the emergence of life." *Geobiology,* 2010.

Sagan, Carl. "Life." *Encyclopaedia Britannica,* 1970.

Shivaji, S., et al. "*Janibacter hoylei* sp. nov., *Bacillus isronensis* sp. nov. and *Bacillus aryabhattai* sp. nov., isolated from cryotubes used for collecting air from the upper atmosphere." *International Journal of Systematic and Evolutionary Microbiology,* 2009.

Steele, Andrew, et al. "Comprehensive imaging and Raman spectroscopy of carbonate globules from Martian meteorite ALH 84001 and a terrestrial analogue from Svalbard" [abstract]. *Geophysical Research Abstracts,* 2007.

———. "Raman imaging spectroscopy of a purported 3.5 billion year old microfossil" [abstract]. *Geophysical Research Abstracts,* 2007.

———. "The science goals of the Astrobiology Field Laboratory (AFL): Report of the MEPAG Science Steering Group." *Astrobiology,* 2007.

Thomas, R. J., S. A. Behnke, et al. "Lightning and electrical activity during the 2009 eruptions of Redoubt Volcano." American Geophysical Union, 2009.

Thomas-Keprta, K., et al. "Origin of Magnetite Nanocrystals in Martian Meteorite ALH84001." *Geochimica et Cosmochimica Act,* 2009.

Toon, Brian, and James Kasting. "A warm, wet Mars?" AbSciCon conference, 2010.

Treiman, A. H. "The history of Allan Hills 84001 revised: Multiple shock events." *Meteoritics & Planetary Science,* 1998.

———. "Submicron magnetite grains and carbon compounds in Martian meteorite ALH84001: Inorganic, abiotic formation by shock and thermal metamorphism." *Astrobiology,* 2003.

Trent, Jonathan, et al. "Possible artefactual basis for apparent bacterial growth at 250°C." *Nature,* 1984.

Vogt, S. S., et al. "Planetary systems with new multiple components." *Astrophysical Journal*, 2007.

Vogt, S. S., P. Butler, et al. "The Lick-Carnegie Exoplanet Survey: A 3.1 M Planet in the Habitable Zone of the Nearby M3V Star Gliese 581." *Astrophysical Journal*, 2010.

Wolfe-Simon, Felisa, et al. "A Bacterium That Can Grow by Using Arsenic Instead of Phosphorus." *Science*, 2010.

INDEX